QD139.P6 PAS

Springer Laboratory

Springer

*Berlin
Heidelberg
New York
Hong Kong
London
Milano
Paris
Tokyo*

Harald Pasch · Wolfgang Schrepp

MALDI–TOF Mass Spectrometry of Synthetic Polymers

With 158 figures and 45 tables

 Springer

Priv.-Doz. Dr. Harald Pasch
Deutsches Kunststoff-Institut
Schloßgartenstraße 6
64289 Darmstadt
Germany
E-mail: hpasch@dki.tu-darmstadt.de

Dr. Wolfgang Schrepp
BASF AG
Kunststofflabor GKP/O-G201
67056 Ludwigshafen
Germany
E-mail: wolfgang.schrepp@basf-ag.de

ISBN 3-540-44259-6 Springer-Verlag Berlin Heidelberg New York

Library of Congress Cataloging-in-Publication-Data applied for

A catalog record for this book is available from the Library of Congress.
Bibliographic information published by Die Deutsche Bibliothek
Die Deutsche Bibliothek lists this publication in the Deutsche Nationalbibliographie; detailed bibliographic data is available in the internet at <http://dnb.ddb.de>.

This work is subject to copyright. All rights are reserved, whether the whole or part of the material is concerned, specifically the rights of translation, reprinting, reuse of illustrations, recitations, broadcasting, reproduction on microfilm or in any other way, and storage in data banks. Duplication of this publication or parts thereof is permitted only under the provisions of the German copyright Law of September 9, 1965, in its current version, and permission for use must always be obtained from Springer-Verlag.
Violations are liable for prosecution under the German Copyright Law.

Springer-Verlag Berlin Heidelberg New York
a member of BertelsmannSpringer Science+Business Media GmbH
http://www.springer.de

© Springer-Verlag Berlin Heidelberg 2003
Printed in Germany

The use of general descriptive names, registered names trademarks, etc. in this publication does not imply, even in the absence of a specific statement, that such names are exempt from the relevant protective laws and regulations and therefore free for general use.

Typesetting: medio Technologies AG, Berlin
Cover design: Künkel & Lopka, Heidelberg
Printed on acid free paper   62/3020/M - 5 4 3 2 1 0

Laboratory Manual Series in Polymer Science

Editors

Prof. Howard G. Barth
DuPont Company
P.O. Box 80228
Wilmington, DE 19880-0228
USA
e-mail: Howard.G.Barth@usa.dupont.com

Priv.-Doz. Dr. Harald Pasch
Deutsches Kunststoff-Institut
Abt. Analytik
Schloßgartenstr. 6
64289 Darmstadt
Germany
e-mail: hpasch@dki.tu-darmstadt.de

Editorial Board

PD Dr. Ingo Alig
Deutsches Kunststoff-Institut
Abt. Physik
Schlossgartenstr. 6
64289 Darmstadt
Germany
email: ialig@dki.tu-darmstadt.de

Prof. Josef Janca
Université de La Rochelle
Pole Sciences et Technologie
Avenue Michel Crépeau
17042 La Rochelle Cedex 01
France
email: jjanca@univ-lr.fr

Prof. W-M. Kulicke
Inst. f. Technische u. Makromol. Chemie
Universität Hamburg
Bundesstr. 45
20146 Hamburg
Germany
email: kulicke@chemie.uni-hamburg.de

Prof. H.W. Siesler
Physikalische Chemie
Universität Essen
Schützenbahn 70
45117 Essen
Germany
email: hw.siesler@uni-essen.de

Preface

Characterization of polymeric materials is vital for predicting and elucidating polymer properties and morphology. Characterization typically involves: (1) molar mass analysis utilizing size exclusion chromatography, light scattering, osmometry, or viscometry, (2) analysis of sequence of repeat units utilizing NMR spectroscopy, (3) endgroup analysis utilizing titration, NMR spectroscopy, or FTIR spectroscopy, and (4) purity examination utilizing NMR spectroscopy, elemental analysis, and FTIR spectroscopy. Until recently, no single technique could completely describe the above characteristics of a polymer sample.

Mass Spectrometry (MS) has been used for the analysis of molar masses of molecules for the past 50 years. However, the application of MS to large biomolecules and synthetic polymers has been limited due to low volatility and thermal instability of these materials. These problems have been overcome to a significant extent through the development of soft ionization techniques such as chemical ionization (CI), secondary-ion mass spectrometry (SIMS), field desorption (FD), fast atom bombardment (FAB), electrospray ionization (ESI) and matrix-assisted laser desorption/ionization time-of-flight mass spectrometry (MALDI-TOF MS). MALDI-TOF mass spectrometry is one of the latest and most fascinating new developments in the analysis of organic compounds. Originally developed for the analysis of biomolecules, it has emerged as one of the most powerful techniques for the characterization of synthetic polymers. It allows for the mass determination of large biomolecules and synthetic polymers of molar masses greater than 1,000,000 Daltons (Da) by ionization and desorption without fragmentation of the analyte molecules. The importance of soft ionization techniques and their far reaching possibilities for the analysis of biomacromolecules has been emphasized just recently by granting the Nobel Prize 2002 for chemistry to John B. Fenn and Koichi Tanaka for their contributions to mass spectrometry.

State-of-the-art MALDI-TOF mass spectrometry is a powerful technique for the fast and accurate determination of a number of polymer characteristics. In particular, it is used efficiently for the fast and accurate determination of molar masses, the sequencing of repeat units, and recognition of polymer additives and impurities. The key feature of the MALDI technique pioneered by Hillenkamp and Karas is the laser irradiation of the sample dispersed in an ultraviolet absorbing matrix. MALDI is a 'soft' ionization technique in which the energy from the laser is spent in volatilizing the matrix rather than in degrading the polymer. Preparation of an appropriate polymer/matrix mixture is one of the critical limiting factors for the general application of MALDI to synthetic polymers. Accordingly, over the past years the challenge has been to discover appropriate matrix materials for use with synthetic polymers.

A number of textbooks on mass spectrometry of polymers have been published in recent years covering the fundamentals of the different techniques. Now, for the first time, a comprehensive presentation of MALDI-TOF mass spectrometry as a powerful analytical technique for synthetic polymers is presented in a book format. In previous publications, MALDI-TOF has been discussed as one technique among others and most emphasis was put on biomolecules instead of synthetic polymers. This book summarizes the fundamentals and technical aspects of the method and describes experimental procedures in detail. It presents all major applications of MALDI-TOF in polymer analysis, e.g., molar mass determination, functional group and copolymer analysis. For the first time, different coupling techniques of MALDI-TOF and liquid chromatography are addressed and important applications are discussed in detail.

The book presents all required information, including instrumentation, selection of matrices and experimental conditions, to enable the reader to plan and execute his own experiments. It includes applications that have not been published before and, thus, presents the latest state-of-the-art. The book is written for beginners as well as for experienced analysts using MALDI-TOF for polymer characterization. It will enable polymer chemists, physicists, and material scientists, as well as students of macromolecular and analytical sciences to optimize experimental conditions for a specific analysis problem. The main benefit for the reader is that a great variety in instrumentation, analysis procedures and applications is given, making it possible to solve simple as well as sophisticated analytical problems.

The authors wish to express their gratitude and appreciation to all colleagues who gave experimental details on specific applications of MALDI-TOF mass spectrometry, in particular Reza Ghahary (Darmstadt), Ralph-Peter Krüger and Steffen Weidner (Berlin), Hans-Joachim Räder (Mainz), Stan Penczek and Andrzej Duda (Lodz), Giorgio Montaudo (Catania), Peter Böshans (Ludwigshafen), Helmuth Kullmann (Ludwigshafen), and Guido Lupa (Ludwigshafen). The support of many colleagues from the Polymer Research Laboratory Ludwigshafen and from German Institute for Polymers (DKI) for providing samples and advice is gratefully acknowledged.

Carefully reviewing a book means lots of work and not much appreciation for the reviewers. Therefore, the authors wish to express their deep gratitude to David Hercules, Nashville, and Joachim Räder, Mainz, for taking on this difficult job.

Finally, the authors thank DKI and the management of BASF Aktiengesellschaft for allowing the publication.

Darmstadt, December 2002 Harald Pasch
Ludwigshafen, December 2002 Wolfgang Schrepp

List of Symbols and Abbreviations

a	Mark-Houwink exponent
ACA	anthracene carboxylic acid
AHB	4-amino-3-hydroxy benzoic acid
AMNP	2-amino-4-methyl-5-nitropyridine
amu	atomic mass unit
ANP	4-amino-5-nitropyridine
AP	aminopyrazine
APCI	atmospheric pressure chemical ionization
ATRP	atom transfer radical polymerization
AUC	analytical ultracentrifuge
A_2	second virial coefficient
BPA	bisphenol A
c_i	concentration of species i
CCA	α-cyano-4-hydroxy cinnamic acid
cf	continuous flow
CI	chemical ionization
CID	collision induced dissociation
CSA	chlorosalicylic acid
D	diffusion coefficient
Da	Dalton
DHB	2,5-Dihydroxy benzoic acid
DLS	dynamic light scattering
DMSO	dimethylsulfoxide
EI	electron impact
EO	ethylene oxide
ESI	electrospray ionization
f	functionality
FAB	fast atom bombardment
FD	field desorption
FFF	field flow fractionation
FT-ICR	Fourier transform ion cyclotron resonance
FTD	functionality type distribution
FTIR	Fourier transform infrared
FT-MS	Fourier transform mass spectrometry
G	Gibbs free energy
GC	gas chromatography

GPC	gel permeation chromatography
GTP	group-transfer polymerization
H	enthalpy
HABA	2-(4-hydroxyphenylazo)benzoic acid
HBM	4-hydroxy-benzylidene malonitrile
HPLC	high performance liquid chromatography
IAA	3,β-indole acrylic acid
k	Boltzmann constant
keV	kilo electron volt
K_{SEC}	distribution coefficient of SEC
K-TFA	potassium trifluoroacetate
kV	kilo volt
LAC	liquid adsorption chromatography
LC	liquid chromatography
LC-CC	liquid chromatography at critical conditions
LD	laser desorption
m	molar mass
MALDI	matrix – assisted laser desorption ionization
MCP	multichannel plate
M_i	molar mass of species i
MM	molar mass
MMD	molar mass distribution
M_n	number-average of molar mass
MPI	multiphoton ionization
MS	mass spectrometry
MSA	5-methoxy salicylic acid
M_v	viscosity-average of molar mass
M_w	weight-average of molar mass
N_A	Avogadros number
NA	9-nitroanthracene
4NA	4-nitroaniline
n_i	molar fraction of species i
NMR	nuclear magnetic resonance
oa	orthogonal acceleration
PA	proton affinity
PαMS	poly-α-methyl styrene
PB	polybutadiene
PBA	polybutylene adipate
PBMA	polybutylmethacrylate
PBSe	polybutylene sebacate
PBSu	polybutylene succinate
PDMS	polydimethylsiloxane
PE	polyethylene
PEG	polyethylene glycol
PEO	polyethylene oxide

List of Symbols and Abbreviations

PI	polyisoprene
PIB	polyisobutylene
PLA	poly-*L*-lactide
PMMA	polymethyl methacrylate
PO	propylene oxide
POM	polyoxymethylene
PPE	polyphenylene ether
PPG	polypropylene glycol
PPO	polypropylene oxide
PS	polystyrene
PSD	post source decay
PVAc	polyvinylacetate
PVP	polyvinylpyridine
Q	polydispersity
QELS	quasi elastic light scattering
R	gas constant
REMPI	resonance enhanced multiphoton ionization
S	entropy
SA	sinapinic acid
S/N	signal to noise ratio
SAN	styrene-acrylonitrile (copolymer)
SBR	styrene-butadiene (copolymer)
SEC	size exclusion chromatography
SIMS	secondary ion mass spectrometry
T	temperature
TDI	toluene diisocyanate
THF	tetrahydrofuran
TOF	time-of-flight
U	acceleration voltage
UV	ultraviolet
v	ion velocity
V_i	interstitial volume
V_o	dead volume
V_R	retention volume
w_i	weight fraction of species i
z	number of charges

Table of Contents

1   **INTRODUCTION** 1
Wolfgang Schrepp

    1.1 Molecular Heterogeneity of Polymers 2
    1.2 Determination of Molar Masses 6
        1.2.1 Light Scattering 7
        1.2.2 Analytical Ultracentrifugation 9
        1.2.3 Viscometry 12
        1.2.4 Size Exclusion Chromatography 13
    References 17

2   **MASS SPECTROMETRIC INSTRUMENTATION** 19
Wolfgang Schrepp

    2.1 Ion Sources 21
        2.1.1 Gas Phase Ionization 22
        2.1.2 Spray Techniques 22
        2.1.3 Ionization from the Condensed Phase 24
            2.1.3.1 Field Desorption (FD) 25
            2.1.3.2 Fast Particles (Fast Atom
                  Bombardment, FAB; Secondary
                  Ion Mass Spectrometry, SIMS) 25
            2.1.3.3 Photons 28
    2.2 Mass Analyzers 29
        2.2.1 Quadrupole Analyzer 29
        2.2.2 Fourier Transform Analyzer 32
        2.2.3 Time-of-Flight Analyzer 34
            2.2.3.1 The Reflectron 36
            2.2.3.2 Time-lag Focusing 37
            2.2.3.3 Practical Consequences 38
    2.3 Detectors 38
    2.4 Data Processing 40
    2.5 Coupling of MS Techniques 42
    2.6 Classical Mass Spectrometry of Polymers 44
        2.6.1 Field- and Laser-Desorption Mass
            Spectrometry 44
        2.6.2 Pyrolysis MS of Polymers 46

	2.6.3 Secondary Ion Mass Spectrometry	48
	2.6.4 Electrospray Ionization (ESI) Mass Spectrometry	50
	References	53

3 FUNDAMENTALS OF MALDI-TOF MASS SPECTROMETRY 57
Wolfgang Schrepp

3.1	The MALDI Process	57
	3.1.1 Primary Ion Formation	59
	3.1.2 Secondary Ionization	61
	3.1.2.1 Gas Phase Proton Transfer	61
	3.1.2.2 Gas Phase Cationization	62
3.2	Matrices and Sample Preparation	62
3.3	Interpretation of Spectra	67
3.4	Experimental Parameters	70
	3.4.1 Influence of the Matrix	70
	3.4.2 Influence of the Cationizing Agent	71
	3.4.3 Influence of the Laser Power	73
	3.4.4 Type of TOF Analyzer	73
	3.4.5 Molar Mass Distribution	78
	References	82

4 IDENTIFICATION OF POLMERS 85
Wolfgang Schrepp

4.1	Introduction	85
4.2	Analysis of Homopolymers	89
	4.2.1 Identification of Low Molar Mass Homopolymers	89
	4.2.2 Polsulfones	91
	4.2.3 Polyester Diols	93
	4.2.4 Poly(oxymethylene)	95
	4.2.5 Emulsifiers	96
	4.2.6 Polysiloxane	99
4.3	Copolymers	100
	4.3.1 Caprolactam-Pyrrolidone Copolymer	100
	4.3.2 Post Source Decay	102
	References	104

5 MOLAR MASS DETERMINATION 107
Harald Pasch

5.1	Introduction	107
5.2	Analysis of Polymers with Narrow Molar Mass Distribution	112

	5.2.1 Oligomer Samples	112
	5.2.2 Hydrocarbon Polymers [12]	115
	5.2.3 Polymers with Varying Polydispersity [5]	121
5.3	Analysis of Polymers with Broad Molar Mass Distribution	123
	5.3.1 Mixtures of Polyethylene Glycols	123
	5.3.2 Mixtures of Polystyrenes and Polymethyl Methacrylates [17]	127
5.4	Further Applications	133
	References	133

6 ANALYSIS OF COMPLEX POLYMERS 135
Harald Pasch

6.1	Introduction	135
6.2	Determination of Endgroups	136
	6.2.1 Polyamides [5]	138
	6.2.2 Triazine-based Polyamines [6]	143
	6.2.3 Poly(methylphenyl silane) [8]	150
	6.2.4 GTP Polymerized PMMA [10]	155
	6.2.5 Poly-ε-Caprolactone [13]	162
	6.2.6 Further Applications	168
6.3	Analysis of Copolymers	168
	6.3.1 Diblock Copolymers of α-Methylstyrene and 4-Vinylpyridine [45]	169
	6.3.2 Copolymers of Ethylene Oxide and Propylene Oxide [47]	176
	6.3.3 Modified Silicone Copolymers [50]	181
	6.3.4 Monitoring the Sulfonation of Polystyrene [52]	185
	6.3.5 Modified Phenol-Formaldehyde Resins [53]	190
	6.3.6 Polyurethanes	199
	6.3.7 Further Applications	203
	References	204

7 COUPLING OF LIQUID CHROMATOGRAPHY AND MALDI-TOF Mass Spectrometry 209
Harald Pasch

7.1	Introduction	209
7.2	Need for LC-MALDI-TOF Coupling	210
7.3	Off-line Measurement of LC Fractions	213
	7.3.1 Calibration of Size Exclusion Chromatography [29]	214
	7.3.2 Molar Mass Distribution of Polyester Copolymers [31, 32]	219

7.3.3 Poly(ethylene oxides) [28]	222
7.3.4 Block Copolymers of Ethylene Oxide and Propylene Oxide	230
7.3.5 Epoxy Resins [37]	233
7.3.6 Block Copolymers of L-Lactide and Ethylene Oxide [38]	245
7.3.7 Further Applications	250
7.4 Interfaces for Coupling LC and MALDI-TOF MS	251
7.4.1 LC Transform Interface	252
7.4.2 PROBOT Interface	254
7.5 Measurement of LC Fractions Using a MALDI-TOF Interface	256
7.5.1 SEC-MALDI-TOF Analysis of Poly(n-butyl methacrylate-*block*-methyl methacrylate) [72]	257
7.5.2 Analysis of Calixarenes by LC-CC-MALDI-TOF [73]	262
7.5.3 Analysis of Polycarbonate by μSEC-MALDI-TOF [76, 77]	266
7.5.4 SEC-MALDI-TOF and LC-CC-MALDI-TOF of Polypropylene Oxides [79]	269
7.6 Direct Coupling of Liquid Chromatography and MALDI-TOF MS	279
References	284
8 RECENT DEVELOPMENTS AND OUTLOOK Wolfgang Schrepp	**289**
References	293
9 SUBJECT INDEX	**295**

1 Introduction

Though both mass spectrometry [1] and the synthesis of polymers [2] seem to be mature techniques – both date back to the beginning of the last century – they have both experienced steady development. The mutual influence of both techniques on each other has been slight but steadily growing and has led to new possibilities, one of which is Matrix-Assisted Laser Desorption/Ionization Mass Spectrometry (MALDI MS).

The development of new ionization techniques like spraying or laser desorption and improvements in the use of Time-of-Flight (TOF) mass analyzers, in detection electronics as well as in the data acquisition, led to an unbelievable increase in attainable mass range and mass resolution. So a rather broad range of polymers, including ionic and polar ones with higher molar masses, became amenable to mass spectroscopic characterization. A rapid commercialization of this new instrumentation coupled with a cost advantage compared to elder equipment rapidly led to a broad distribution within the polymer community.

The content of this book will be as follows: after a short discussion of the interest in molar mass characterization in general, an overview of other techniques for molar mass characterization will be given including mass spectrometric techniques different from MALDI MS. A description of the main tasks of a mass spectrometer – ion generation, ion separation, ion detection – will introduce to the fundamentals of the MALDI process. The introductory analysis of MALDI spectra of standard polymers will be followed by an in-depth description of the analysis of complex polymer structures. The advantages of using pre-separation techniques like size exclusion (SEC) or multidimensional chromatography, including the application of MALDI-TOF MS as a detector, will be highlighted. Future developments will be outlined.

The idea of this publication is twofold: first an up-to-date overview of the possibilities and limits of the MALDI technique for synthetic polymers, excluding the rapidly growing field of biopolymers, shall be given. In addition, practical guidance for beginners and practitioners to perform initial experiments shall be presented by describing detailed examples. The intention is not to give a comprehensive overview of mass spectrometry of synthetic polymers.

1.1 Molecular Heterogeneity of Polymers

Molar mass is one of the most important parameters characterizing polymer properties, like mechanical behaviour, melting temperature, and the viscosity of melts. Accordingly, the molar mass (distribution) is one of the key parameters determining the range of applications of a polymer. An example for low molar mass polyethylenes (PEs) adapted from Ehrenstein [3] is given in Table 1.1.

Up to a molar mass of 20,000 g/mol many polymers are of waxy character [3]. Technical polymers used as films or structural materials often show molar masses up to 1 million g/mol. An overview of physical parameters depending on the molar mass for high-density PE is given in Fig. 1.1 adapted from [3].

Not only the molar mass itself but also the molar mass distribution plays a significant role. So the toughness of polymers with

Table 1.1. Properties of low molar mass polyethylenes at 23 °C

M_w (g/mol)	Tensile strength (N/mm^2)	Density (g/cm^3)	Phase
1400–10,000	3–10	0.92–0.96	Solid
250–1400	2	0.87–0.93	Solid
70–240	–	0.63–0.78	Liquid

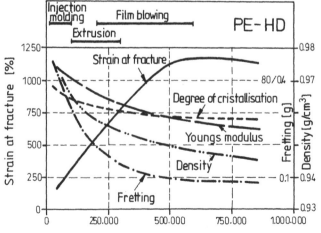

Fig. 1.1. Dependence of physical properties on molar mass for polyethylene. (Reprinted from [3] with permission of Carl Hanser, Germany)

1.1 Molecular Heterogeneity of Polymers

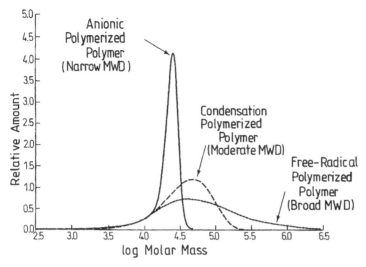

Fig. 1.2. Influence of the type of polymerization reaction on the molar mass distribution. (Reprinted from [4] with permission of Springer, Berlin Heidelberg New York)

a narrow molar mass distribution produced by special catalysts like metallocenes may be considerably higher than for conventional materials.

Due to the statistical nature of polymerization reactions, polymers consist of a mixture of molecules comprising a certain range of molar masses. From a mass spectroscopist's view even the simplest polymer consists of a more or less complex mixture of molecular species. This fact constitutes a remarkable difference to other samples investigated by mass spectrometry like pharmaceuticals or monomers, and surely causes some difficulties in the application of mass spectrometry to polymers. The width of the molar mass distribution (MMD) depends on the polymerization mechanism, the kinetics, and the reaction conditions (see Fig. 1.2).

The distribution with respect to molar mass represents only one type of molecular complexity. As visualised in Fig. 1.3, besides the MMD other types of heterogeneity come into play. The use of different monomers leads to copolymers (block-, graft-, statistical), that is to chemical heterogeneity (CH). Polymers can be linear, branched (comb- or star-like), or cyclic, that is they can show different molecular architectures (MA). Depending on the polymerization procedure the macromolecules can contain functional groups at their chain ends (telechelics or macromonomers) leading to a functionality type distribution (FTD).

In addition, the different types of molecular heterogeneity may superimpose on each other. There is no single omnipotent method

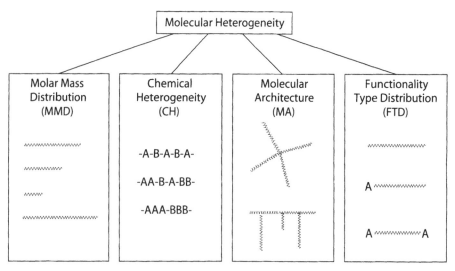

Fig. 1.3. Schematic representation of possible molecular heterogeneities of polymers (Reprinted from Ref. [5] with permission of Springer, Berlin Heidelberg New York)

available giving, simultaneously, information on all these types of heterogeneity, but we will see that making proper use of MALDI MS together with suitable pre-separation techniques at least part of the molecular complexity can be analysed (MMD, CH, and FTD).

Depending on the physical quantity encountered and the determination method used, different average values of the molar mass of a polymer have to be considered. These averages are defined in terms of the molar mass M_i of species i, their corresponding number n_i, weight w_i or concentration c_i. The number-average molar mass M_n is defined as

$$M_n = (\Sigma n_i M_i)/(\Sigma n_i) \qquad (1.1)$$

M_n is determined by (colligative) methods sensitive to the number of molecules like osmometry or cryoscopy. The weight-average molar mass M_w is defined as

$$M_w = (\Sigma n_i M_i^2)/(\Sigma n_i M_i) = (\Sigma w_i M_i)/(\Sigma w_i) = (\Sigma c_i M_i)/(\Sigma c_i) \quad (1.2)$$

M_w is obtained by methods sensitive to the weight of molecules like sedimentation or light scattering. M_n and M_w have a unique value for a given polymer sample.

Another molar mass average of interest is the value obtained by viscosity measurements, for which temperature and solvent are of importance. The definition is

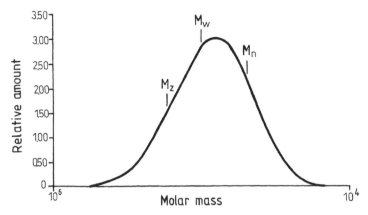

Fig. 1.4. Illustration of the typical average values of a molar mass distribution

$$M_v = [(\Sigma w_i M_i^a)/(\Sigma w_i)]^{1/a} \quad (1.3)$$

The exponent a is obtained by the Mark-Houwink relationship. The value of a is between 0.5 and 1 for random coils and approaches 1.8 for rigid rod molecules. With these limits for a it can be seen that M_v is always larger than M_n but can be equal to M_w for the upper limit of random coils ($M_w \geq M_v \geq M_n$). Finally, the so-called z-average of molar mass shall be mentioned. It is determined by sedimentation equilibrium measurements and is defined as

$$M_z = (\Sigma n_i M_i^3)/(\Sigma n_i M_i^2) \quad (1.4)$$

Due to the weighing factor $n_i M_i^2$ higher molar masses are accentuated even more as compared to M_w values. A scheme illustrating the various averages of molar mass is given in Fig. 1.4.

A measure of the width of the molar mass distribution is the ratio of the average values: the polydispersity Q is defined as

$$Q = M_w/M_n \quad (1.5)$$

Q equals 1 for monodisperse samples, meaning all molecules have the same molar mass. Many polycondensation products show values of 2; technical materials usually have considerably higher polydispersities (up to 20).

In contrast to the determination of average values of the molar mass, the determination of functional endgroups, e.g., by infrared spectroscopy or nuclear magnetic resonance, is tedious due to the low concentration of these groups. It will be shown that MALDI MS is a useful means for solving this problem and in combination with special fractionation techniques can provide information even on the functionality type distribution of a polymer (see Chap. 6).

1.2 Determination of Molar Masses

In this section a short description of the most widely used methods for molar mass determination of synthetic polymers apart from mass spectrometry is presented. A classification of these methods with respect to mass range, average value determined and type of measurement, absolute method or relative method requiring calibration, is given in Table 1.2. This overview should allow a better judgement of the potential of mass spectroscopic methods.

The methods indicated in bold type will be described in more detail, to give at least one example of how to determine the various mean values and to show posibilities and limitations of some other important molar mass methods.

Table 1.2. Important methods for molar mass determination with indication of the attainable range

Average	Methods	Molar mass range (g/mol)	Type
M_n	Ebullioscopy, cryoscopy,		
	vapor pressure osmometry,	$<10^4$	a
	membrane osmometry,	$5 \cdot 10^3 - 10^6$	a
	sedimentation equilibrium,	$10^2 - 10^6$	a
	field flow fractionation,	$10^4 - 10^7$	r
	size exclusion chromatography	$10^3 - 10^7$	r
M_w	Static *light scattering,*	$10^4 - 10^7$	a
	field flow fractionation,	$10^4 - 10^7$	r
	size exclusion chromatography	$10^3 - 10^7$	r
M_z	*Sedimentation equilibrium,*	$10^2 - 10^7$	a
	Size exclusion chromtography	$10^3 - 10^7$	r
M_v	*Viscosity measurement*	$>10^2$	r

a = absolute method; r = relative method

1.2.1 Light Scattering

Light scattering of dissolved polymer molecules [6,7] relies on the fact that incident light waves cause a mutual shift of the negative and positive charges of matter. The induced dipoles emit an electromagnetic wave of the same frequency as the incident light. This radiation is coherent and elastic, and is called Rayleigh scattering. Apart from the elastic scattered light, inelastic contributions due to particle diffusion (Doppler effect) or density fluctuations caused by sound waves (Brillouin scattering) can be observed.

Using the theoretical considerations of Debye and Rayleigh for the coherent scattering of light, Zimm in 1948 [8] derived the fundamental equation used in static light scattering:

$$(K \cdot c_2)/R(\theta) = 1/M_2 + 2A_2 c_2 + \ldots \qquad (1.6)$$

R and K are defined as $R(\theta) \equiv (r^2(I_s/I_0)/(V(1 + (\cos\theta)^2)))$ and $K \equiv 2\pi^2 n_0^2 (dn/dc)^2/N_A \lambda_0^4$ with r the distance that the light travelled within the medium, V the scattering volume, n the refractive index, dn/dc the refractive index increment, c_2 the concentration of the solute, M_2 the molar mass of the solute, λ_0 the wavelength of the incident light, I_0 the incident light intensity, I_s the scattered intensity of the incident radiation, N_A Avogadros number, A_2 the second virial coefficient, and θ the angle between the direction of the incident light wave and the observer.

In scattering experiments it is common to use the so-called scattering vector which is defined as $q = (4\pi/\lambda_0) \sin(\theta/2)$. The term $R(\theta)$ can be expressed as function of q so that one can write $R(q)$ instead of $R(\theta)$.

Assuming that the macromolecules are chemically uniform and differ only in molar mass, for very dilute solutions for the component i with concentration c_i the following expression is obtained:

$$R(q) = \Sigma_i R(q)_i = \Sigma_i K M_i c_i \qquad (1.7)$$

For $c = \Sigma_i c_i$ one obtains

$$Kc/R(q) = \Sigma_i c_i / \Sigma_i c_i M_i = 1/M_w \qquad (1.8)$$

i.e., static light scattering determines the weight average of molar mass by following the coherent, elastic part of scattering.

Two common experimental arrangements for performing light scattering experiments are given in Fig. 1.5. Fig. 1.5A is a schematic representation of the set-up for wide angle scattering. A la-

Wide Angle Light Scattering

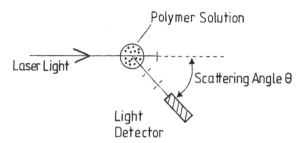

Small Angle Light Scattering (θ=6°)

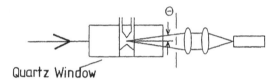

Fig. 1.5. Experimental arrangements for static light scattering: wide angle light scattering and small angle light scattering

ser beam (typically 633 or 532 nm) passes through a dilute polymer solution. The scattered light intensity I_s (in general only 10^{-4} to 10^{-5} of the incident intensity I_0) is monitored over a wide range of angles for a series of polymer concentrations c. The data are plotted in a so-called Zimm diagram (see Fig. 1.6) with Kc + $\sin^2(\theta/2)$ on the abscissa and Kc/I_s on the ordinate axis. Extrapolation to c=0 and θ=0 gives $1/M_w$. This plot yields two additional molecular parameters, the second virial coefficient A_2 and R_g, and the z average of the radius of gyration. The wide angle measurement is typically applicable within a molar mass range $10,000 < M_w < 10,000,000$ g/mol. Due to the weak intensity of the scattered light, great care has to be taken to prevent the presence of dust particles.

In small angle light scattering which monitors only light typically scattered into an angle range between 6° and 7°, the extrapolation to θ=0 can be omitted; the extrapolation to c=0 again delivers $1/M_w$ and A_2. The advantage of this arrangement is higher sensitivity allowing the study of molecules of smaller mass; a dis-

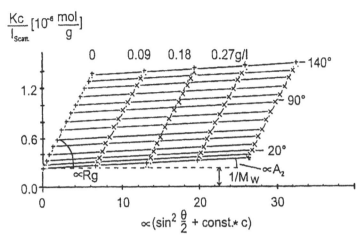

Fig. 1.6. Wide angle light scattering of a high molar mass cationic polyelectrolyte poly(acrylamide/dimethylaminoethylacrylate)*CH_3Cl in 0.5 mol/l NaCl; M_w = 4.2×10^6 Da, R_g = 153 nm, A_2 = 3×10^{-4} mol ml/g²)

advantage is the lack of information on the conformation of the molecules expressed through R_g.

In dynamic or quasielastic light scattering (DLS or QELS), light intensity fluctuations induced by the Brownian motion of the particles or macromolecules are analyzed by evaluating the intensity time correlation function. This yields the diffusion coefficient in the given solvent. By means of the Stokes-Einstein equation and the viscosity of the solvent the hydrodynamic radius of the species under investigation is obtained. Thus, in combination with static light scattering, the architecture of the macromolecules can be obtained [7,9].

Light scattering in practical use is limited to special applications due to the fact that it requires freedom from dust with the associated filtration problems, gives only weight averages of molar mass, and its application is tedious with multimodal distributions. For the general determination of polymer molar masses chromatographic methods are more widespread.

1.2.2 Analytical Ultracentrifugation

The principle of ultracentrifugation is to expose a solution or dispersion to a high centrifugal field and to detect the resulting shift in concentration by an absorption or refractive index (RI) detector. The history of molar mass determination is closely linked to the development of the ultracentrifuge. In 1925 Svedberg and Fahrens [10] performed the first definite high molar mass measurement, determining 68,000 g/mol for hemoglobin. Differences between M_n and M_w were first pointed out for polyesters by Krämer and Lansing [11], Signer and Gross [12], and McCormick [13].

The most common methods for the analytical ultracentrifuge (AUC) are: the sedimentation velocity run, in which fractionation according to size takes place, the sedimentation equilibrium run, and the density gradient run, which fractionates according to the density of the dissolved species (chemical heterogeneity, degree of grafting, number of components in a mixture). The strength of the AUC lies in the fact that it can determine all relevant features of an MMD (M_n, M_w, M_z) and especially the capability to analyze dissolved macromolecules and dispersed microparticles simultaneously.

When the rate of sedimentation, which depends on the mass and shape of the dissolved species, and the viscosity of the solvent are balanced by the rate of diffusion, the system is in a state of equilibrium sedimentation. In the ultracentrifuge a molecule (or particle) experiences three types of forces:

(1) the force caused by the radial acceleration given by Newtons second law

$$F_a = M\omega^2 r / N_A \qquad (1.9)$$

with r being the distance of the axis of rotation and ω being the angular velocity;

(2 and 3) the two counteracting forces are the buoyant force and the viscous resistance. The former is expressed as

$$F_b = V\varrho\omega^2 r = M/(N_A \varrho_2) \cdot \varrho\omega^2 r \qquad (1.10)$$

with ϱ and ϱ_2 being the densities of the solution and the solute, respectively, V being the volume of particle. The viscous force can be expressed as

$$F_v = f\, v_s \qquad (1.11)$$

with the friction coefficient f. In equilibrium ($F_a = F_b + F_v$) and with $v_s = dr/dt$:

$$dr/dt = M/(fN_A) \cdot (1 - \varrho/\varrho_2)\omega^2 r \qquad (1.12)$$

The stationary state velocity per unit acceleration is called the sedimentation coefficient s where

$$s = (dr/dt)/(\omega^2 r) = M/(fN_A) \cdot (1 - \varrho/\varrho_2) \qquad (1.13)$$

Many biopolymers, for example, are simply identified in terms of their sedimentation coefficient instead of molar mass.

If a sedimentation experiment is carried out for a sufficiently long time, a state of equilibrium is reached between sedimentation and diffusion. The two counteracting fluxes (driven by centrifugal field and concentration gradient) are

$$J_{sed} = cv_s = c\omega^2 rs \qquad (1.14)$$

and the flux due to diffusion by Ficks first law, D being the diffusion coefficient:

$$J_{diff} = -Ddc/dr \qquad (1.15)$$

In the sedimentation equilibrium the sum of the two fluxes must be zero resulting in

$$Ddc/dr = c\omega^2 sr \qquad (1.16)$$

which gives after integration

$$\ln c = s/(2D)\cdot\omega^2 r_2 + \text{const.} \qquad (1.17)$$

With the definition for s and $D = kT/f$ we have

$$\ln c = M(1-\varrho/\varrho_2)/(2N_A kT)\cdot\omega^2 r^2 + \text{const.} \qquad (1.18)$$

This expression is equivalent to the barometric formula which describes the variation of atmospheric pressure with height.

It is clear from Eq. (1.18) that a measurement of the solute concentration at different values of r and plotting $\ln c$ over r^2 gives the slope $M(1-\varrho/\varrho_2)/2RT$ which allows the determination of M. A more in-depth consideration given in the literature [7,14,15] reveals that AUC is a comprehensive method for the determination of various molar mass averages, the MMD, as well as sedimentation and diffusion coefficients. As Eq. (1.18) is valid for ideal solutions, the molar mass value M_w is obtained by plotting 1/M as a function of c and extrapolating c to zero.

The AUC also yields particle size distributions of latices and dispersions and the content of macromolecules in a serum phase. In the so-called density gradient run information on chemical heterogeneity, number of components in a mixture, and degree of grafting can be obtained.

1.2.3 Viscometry

As was already pointed out in Sect. 1.1, the viscosity of a polymer solution is directly correlated to the molar mass of the solute. Like in AUC the determination of molar mass by measuring the viscosity is connected with the early days of polymer science as the first measurements of cellulose and cellulose derivatives by Staudinger and Freudenberger [16] played an important role in establishing the concept of macromolecules. Staudinger also proposed a direct proportionality between the limiting viscosity and the molar mass which, after a semiempirical modification by Mark [17], Houwink [18], and Sakurada [19], became the form used today:

$$[\eta] = kM^a \qquad (1.19)$$

with $[\eta] = \lim_{c \to 0} \eta_{sp}/c = \lim_{c \to 0}((\eta-\eta_0)/\eta_0)/c$ and η being the viscosity of the polymer solution, η_0 being the viscosity of the solvent. The constants k and a are called Mark-Houwink coefficients for a given system. The numerical values depend on both the polymer and the solvent at constant temperature. Extensive tabulations of k and a are available. The exponent a can be understood as a parameter describing the conformation of the macromolecules. It ranges from 0 for spherical molecules (no dependency of the intrinsic viscosity on molar mass) to 2 for rigid rod molecules. For flexible molecules a ranges from 0.5 to 0.8.

The use of Eq. (1.19) necessitates the extrapolation of η_{sp}/c to $c=0$. Various empirical approximations are available, for example the one given by Huggins [20]:

$$\eta_{sp}/c = [\eta] + [\eta]^2 k_H c + \ldots \qquad (1.20)$$

By plotting η_{sp}/c over c and assuming that the linear relationship holds even in the very dilute regime, $[\eta]$ can be obtained by extrapolation to $c=0$.

Equation (1.19) is valid only for monodisperse polymer chains (i) with molar mass M_i:

$$[\eta]_i = k_i M_i^a \qquad (1.21)$$

When a polydisperse sample contains the monodisperse part with molar mass M_i in the weight ratio $W(M_i)$, then the overall $[\eta]$ is given by the superposition of $[\eta]_i$ as follows:

$$[\eta] = k' \Sigma_i M_i^a W(M_i) \qquad (1.22)$$

with k' being the k-value for monodisperse polymers. The above equation holds because of the experimental finding that the relative viscosity η_{sp} is additive. With the definition of the viscosity average of molar mass being

$$M_v = (\Sigma_i M_i^a W(M_i))^{1/a} \qquad (1.23)$$

one obtains

$$[\eta] = k' M_v^a \qquad (1.24)$$

Generally the viscosity of polymer solutions is measured in capillary or rotational viscometers. As viscometry needs calibration and delivers a special type of molar mass average, it is mainly used if a particular polymer has to be characterized routinely.

1.2.4 Size Exclusion Chromatography

Size exclusion chromatography (SEC) [4] is the most popular and convenient method for the molar mass characterization of polymers. Typically in less than 30 min the complete MMD of a polymer together with all statistical moments of the distribution can be obtained – in notable contrast to the methods described so far.

The separation process is driven by the molecular hydrodynamic volume of the polymer species. The polymer is dissolved in a suitable solvent and passed through a column packed with porous particles. Higher molar mass species that are too large to penetrate into the pores elute first; smaller ones that can diffuse into the pores appear at higher elution volumes. The SEC system has to be calibrated with a series of solutes of known molar mass. As can be inferred from Fig. 1.7, a relationship between elution volume and log M is established.

The high molar mass solutes which do not penetrate the pores of the packing material elute within the interstitial or void volume V_0 of the column, i.e., the volume of the mobile phase located between the packing particles. With lower molar mass the molecules will partition or penetrate into the pores and elute at longer elution times. Finally, if the molecules become so small that they diffuse freely in the pores, they sample the total pore volume V_i of the packing. So the elution volume of the small molecules is equal to the volume of the total mobile phase of the column $V_t = V_0 + V_i$. The retention volume V_R of an arbitrary molecule between these two limiting cases is described as

$$V_R = V_0 + K_{SEC} V_i \qquad (1.26)$$

with K_{SEC} being the SEC distribution coefficient, which is defined as the ratio of the average concentration of the solute in the pore

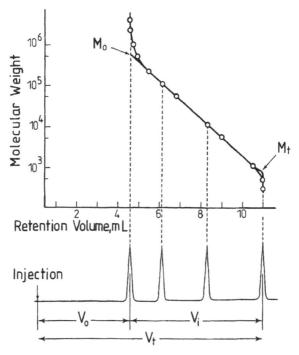

Fig. 1.7. SEC calibration curve as a plot of log molar mass vs elution or retention volume of a series of polymer standards. The total permeation volume of the column V_t equals the sum of the exclusion or void volume V_0 and the pore volume of the column packing V_i. The extrapolated value M_0 defines the exclusion limit, M_t the total permeation limit of the column. (Reprinted from [4] with permission of Springer, Berlin Heidelberg New York)

volume to that in the interstitial volume. K_{SEC} can vary between 0 ($V_R = V_0$; molecules too large to penetrate the pores; exclusion limit) and 1 ($V_R = V_0 + V_i = V_t$; all the pores are accessible to the analyte molecules, no change in conformational entropy; separation threshold).

The SEC process can be described by thermodynamic considerations. In general, the partition coefficient K_d betweeen the mobile and the stationary phase is related to the change in Gibbs free energy G in the usual way through enthalpy H, entropy S, and temperature T:

$$\Delta G = \Delta H - T\Delta S = -RT\ln K_d \qquad (1.27)$$

$$K_d = e^{(\Delta S/R - \Delta H/RT)} \qquad (1.28)$$

In ideal SEC, separation is exclusively directed by conformational changes, i.e., no interaction with the packing material ($\Delta H=0$) takes place:

$$K_{SEC} = e^{\Delta S/R} \qquad (1.29)$$

This is in contrast to another mode of liquid chromatography described in Chap. 7 which makes use of adsorptive interactions between the macromolecules and the stationary phase, being called liquid adsorption chromatography (LAC). Here ΔH is maximized and the separation occurs according to the chemical compositon of the macromolecules.

SEC measurements can be carried out using a set-up consisting of a solvent delivery system (solvent reservoir, pump, degasser), the separation system (injector, thermostated SEC column), and a detection system. Detectors are used to monitor concentration changes in the column effluent. They can be classified into three categories: concentration-based (differential refractometer, evaporative light scattering detector), structure-sensitive (UV- or IR-detector), and molar mass-sensitive detectors (viscosity, light scattering, mass spectrometry). This last point, representing one of the valuable features of MALDI MS, will be described in detail in Chap. 7.

Currently many manufacturers offer data handling systems which include software for SEC calculations. For example, by injecting PS standards, calibration curve equations can be generated automatically as well as the calculation of molar mass averages and the MMD.

For a better understanding of the principle a short simplified description of the procedure will be outlined. As in the first stage, SEC delivers a diagram of the type "arbitrary intensity vs retention volume"; this chromatogram, by using a calibration curve, has to be converted into the corresponding molar mass distribution from which the statistical moments can be calculated. An illustration is given below; see Fig. 1.8. Here the chromatogram is divided into i equally spaced retention volumes with height h_i and the corresponding molar mass M_i through the calibration. The molar mass averages now can be calculated as follows:

$$M_n = \Sigma h_i/(\Sigma h_i/M_i), M_w = \Sigma h_i M_i/\Sigma h_i, M_z = \Sigma h_i M_i^2/(\Sigma h_i M_i) \quad (1.30)$$

Usually the MMD from SEC is given as an arbitrary intensity vs log molar mass curve; see Fig. 1.9. Principally weak points of SEC are the fact that it needs calibration, which for more complex polymers is a problem due to the lack of standards, insufficient resolution for oligomers, and determination of cyclic species. In the case of complex polymers like copolymers, for example, the application of mass-sensitive detectors further enhances the information content of SEC runs. So by using multiangle light scattering detectors

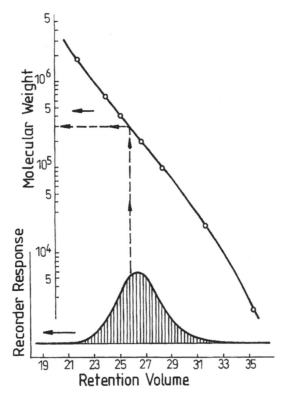

Fig. 1.8. SEC chromatogram of a PS standard (NBS 706) and the corresponding calibration curve. (Reprinted from [4] with permission of Springer, Berlin Heidelberg New York)

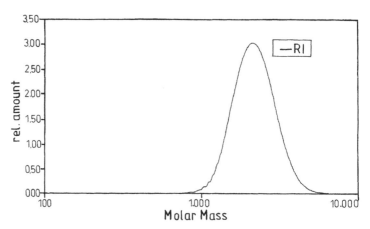

Fig. 1.9. Typical SEC result (PMMA, M_w 2400 g/mol) giving the envelope of the oligomer distribution

the distribution of the radius of gyration can be obtained. The same holds true for the application of viscosity detectors which in addition can give the M_n of copolymers.

In conclusion, the preceeding sections show that there is a variety of methods available for the determination of molar masses delivering more or less information, even on the MMD. Thus the question may arise: what is the need for an additional method in this area? As has been shown, many of the methods described so far need a calibration, i.e., they do not deliver absolute molar mass information. There are restrictions in the molar mass range detectable, especially with respect to lower masses. The experimental procedures might be tedious and due to the limited resolution little information on the chemical heterogeneity is available. Precisely this information can be provided by mass spectroscopic techniques.

References
1. SVEC HJ (1985) Int J Mass Spectrom Ion Proc 66:3
2. ELIAS H-G (1999) Macromolecules. Verlag Hüthig und Wepf, Basel
3. EHRENSTEIN GW (1999) Polymer-Werkstoffe. Hanser Verlag, München
4. MORI S, BARTH HG (1999) Size exclusion chromatography. Springer, Berlin Heidelberg New York
5. PASCH H (1997) Adv Polym Sci 128:1
6. HUGLIN MB (ed) (1972) Light scattering from polymer solutions. Academic Press, New York
7. LECHNER MD, GEHRKE K, NORDMEIER EH (1993) Makromolekulare Chemie. Birkhäuser, Basel
8. ZIMM BH (1948) J Chem Phys 16:1099
9. HIEMENZ PC (1986) Principles of colloid and surface chemistry. Marcel Dekker, New York
10. SVEDBERG T, FAHRENS R (1926) J Am Chem Soc 48:430
11. KRAEMER EO, LANSING WD (1935) J Phys Chem 39:153
12. SIGNER R, GROSS H (1934) Helv Chim Acta 17:59, 335, 726
13. MCCORMICK HW (1959) J Polym Sci 41:327
14. COOPER AR (ed) (1989) Determination of molecular weight. Wiley, New York
15. MÄCHTLE W, HARDING S (eds) (1991) AUC in biochemistry and polymer science. Cambridge, UK
16. STAUDINGER H, FREUDENBERGER H (1930) Ber Dtsch Chem Ges 63:2331
17. MARK H (1983) Der feste Körper. Hirzel, Leipzig
18. HOUWINK R (1940) J Prakt Chem 155:241
19. SAKURADA I (1941) Kasenkouenshyu 6:177
20. HUGGINS ML (1942) J Am Chem Soc 64:2716

2 Mass Spectrometric Instrumentation

This chapter describes the basic principles of mass spectrometry to allow a better assessment of the MALDI-technique. The typical procedure in mass spectrometry is as follows. After introducing the sample into the mass spectrometer, in the first step ions characteristic for the sample under investigation have to be created, i.e., an ion source is needed. Then the ions have to be separated according to their mass-to-charge ratio (m/z) which is the task of the (mass) analyzer. In the end the separated ions have to be detected and the signal displayed in a convenient form. Thus a mass spectrometer consists of the building blocks shown in Fig. 2.1.

In mass spectrometry, which separates ions according to their mass and charge, it is usual to give the masses in atomic mass units (amu); 1 amu is defined as 1/12 of the mass of the ^{12}C atom. As the mass number of an isotope gives the number of the constituting protons and neutrons and the mass of these elementary particles is about 1 amu, it is in many cases with sufficient accuracy possible to use this unit instead of the atomic mass (for example ^1H 1.007825 amu, ^{16}O 15.994915 amu). By rounding to full numbers one obtains the so-called nominal mass for which also the sign Dalton (Da) is in use (the ion of benzene $^{12}C_6{}^1H_6$ gives 78 Da; 1 Da designates the mass of a hypothetical atom with atomic mass 1). For the exact calculation of masses it is not possible to use the atomic masses which give the mean value of the natural isotope mixture (35.45 for chlorine for example which consists of isotopes 35 and 37); instead the isotope masses have to be used. In many cases only singly charged ions are observed, that is m/z value and mass number are numerically equal.

Two parameters frequently used for the characterization of the performance of mass spectrometers are *mass resolution* and *sensitivity*. Mass resolution describes the capability of a mass spec-

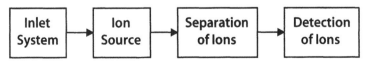

Fig. 2.1. Principle of a mass spectrometer

trometer to separate ions of adjacent mass numbers m and m + Δm. The resolution is given by

$$R = m/\Delta m \quad (2.1)$$

or in parts per million

$$R(ppm) = 10^6 \, \Delta m/m \quad (2.2)$$

Two peaks are regarded as resolved if (h/H)×100 ≤ 10 where H is the height of the peak and h is the height of the valley between them; see Fig. 2.2.

If the masses 100.00 and 100.01 are separated by a 10% valley the resolution is 10,000. Resolution in the mass range relevant to polymers can be tested by separating the natural isotope pattern of C- and H-containing compounds (standard polymers like PEG, PS, or PMMA can be used). Magnetic sector instruments operate with a constant resolution over a suitable mass range, for example 10,000 from 300 to 1000 amu. Quadrupole instruments are often operated at a constant Δm giving for Δm = 1 resolutions of 1000 at 1000 amu and 300 at 300 amu.

Sensitivity can be tested by determining the minimum amount of sample consumption to obtain a useful signal. In MALDI MS the definition of a certain signal-to-noise ratio (S/N) for a given amount of material is used, for example S/N of 30:1 for 100 fmole of angiotensin (a protein) at 1047 amu.

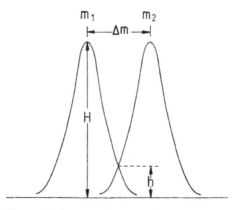

Fig. 2.2. Explanation of mass resolution. (Reprinted from [1] with permission of University Press Cambridge, UK)

2.1 Ion Sources

A first important step in mass spectrometry is the generation of ions from the sample of interest. Depending on the type of sample different ionization techniques are used. An overview on the various possibilities is given in Table 2.1.

In general, ions or excited species are formed by the removal or addition of an electron giving a radical cation $M^{+\bullet}$ or anion $M^{-\bullet}$, respectively, or by addition of other charged species:

$$M + e^- \rightarrow M^{+\bullet} + 2e^- \qquad (2.3)$$

and with lower probability

$$M + e^- \rightarrow M^{-\bullet} \qquad (2.4)$$

or

$$M + e^- \rightarrow M^* + e^- \qquad (2.5)$$

or

$$M + X^+ \rightarrow [M + X]^+ \qquad (2.6)$$

where M denotes the molecular species to be ionized, e^- an electron; the * indicates an excited species and X may be a proton or as in MALDI a Na-, K-, Li- or Ag-atom.

Depending on the analyte structure, the thermal energy prior to ionization, the energy gained during ionization, and the environment of the ion, determining whether it collides or not with other species, further fragmentation can happen leading to other ions A and neutrals N:

$$M^{+\bullet} \rightarrow A^+ + N^{\bullet} \text{ or } A^{+\bullet} + N \qquad (2.7)$$

Fragmentation can lead to complications of the mass spectra but on the other hand can be used as an assignment aid if it is made to

Table 2.1. Different ways of producing ions

Ionization agent	Acronym of ion source
Electrons	Electron impact (EI)
Electric fields	Field desorption (FD); field ionization (FI)
Electric fields and spraying	Electrospray ionization (ESI) (or thermospray)
Rapidly moving ions	Secondary ion mass spectrometry (SIMS)
Photons	Laser desorption (LD); multiphoton ionization (MPI); matrix-assisted laser desorption/ionization (MALDI)

occur in a controlled manner. The next sections will deal with three types of ionization: gas phase ionization, spray techniques, and ionization in condensed phase.

2.1.1 Gas Phase Ionization

Volatized molecules can be ionized most simply by a beam of electrons from a hot wire filament. The electron impact ion (EI) sources are historically among the first ion sources developed [2] and are still widely used today. Reasons for this popularity are stability, ease of operation, relatively high sensitivity, and the fact that comprehensive libraries exist on the EI-basis. Standard mass spectra are obtained conventionally at an acceleration voltage of 70V across the ion chamber, as the ionization cross-section is at maximum between 50 and 100eV for most molecules and the fragmentation pattern is strongly dependent on the electron energy. The desired outcome due to interaction of the electron beam and the molecules is electron abstraction leading to

$$M + e^- \rightarrow M^{+\bullet} + 2e^- \tag{2.8}$$

However, other processes like electronic excitation or radical anion formation are also possible. The ionization following Eq. 2.8 leads to the formation of molecular ions with a wide range of internal excess energies and hence a considerable amount of fragmentation. As electron capture is more than a factor of 100 less probable (depending on the molecular structure of the analyte) than electron removal, negative-ion mass spectrometry by EI is inherently less sensitive than positive-ion spectrometry. As not all molecules react with the electron beam and those which interact can produce a variety of products, probably (much) less than 1% of the sample molecules are converted into detectable ions.

The main obstacle in using EI sources for the investigation of polymers is the fact that the species have to be brought into the gas phase without fragmentation which is difficult if not impossible for most polymers. Therefore EI sources, despite their widespread use in other areas, only play a marginal role in the analysis of polymers.

2.1.2 Spray Techniques

Many polymers are soluble in organic solvents or even water, for example biopolymers or surfactants, so that techniques for the desorption of ions from liquids might be a proper means for generating mass spectra. The principle steps in spray ionization techniques which allow the production of ions from liquids are:

1. Generation of small droplets
2. Shrinkage of the droplets and
3. Ion formation in the gaseous phase

Disruption of the liquid necessitates an energy source which can be of aerodynamic, thermal, or electrical nature. The case of electrospray ionization (ESI) [3] will be considered in more detail as it is a technique showing possibilities for the analysis of oligomers and even polymers. The technique was first conceived in the 1960s by Dole et al. [4] and brought into practice in the 1980s by Fenn [5]. Who was honored in 2002 by the Nobel price for chemistry.

In electrospray ionization the liquid is forced through a capillary tube to which a high voltage (some kV) is applied with a counterelectrode located some distance away from the capillary; see Fig. 2.3. Evaporation of the solvent from the initially formed droplets as they pass through a pressure gradient towards the analyzer leads to a shrinkage of the multiply charged droplets. This process is enhanced through the use of a dry and warm sheath gas. At the point where the magnitude of the charge repulsion is sufficient to overcome the surface tension holding the droplet together, desolvated ions are formed. ESI sources are generally operated at atmospheric pressure and temperatures close to ambient. These conditions favor a multiple charging of the ions. Mole-

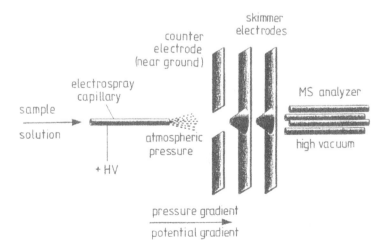

Fig. 2.3. Scheme of the electrospray set-up. (Reprinted from [3] with permission of Wiley, UK)

cules that possess various sites being capable of protonation (i.e., basic NH_2-groups in proteins) or deprotonation can be evaporated as quasi-molecular ions $[M+zH]^{z+}$ or $[M-zH]^{z-}$, respectively. The choice between positive- or negative-ion operation depends on the nature of the sample.

The beneficial effect of increasing the charge well above unity is to bring molecules with larger masses m into the m/z range of simpler mass spectrometers like quadrupole systems. Another significant point is that ESI provides a logical coupling of solution introduction and the ionization of highly polar and involatile compounds. So a direct coupling to other separation techniques like high performance liquid chromatography (HPLC) is possible.

The electrospray conditions imply that the ions have to be present already in the solution either because the analyte is ionic (i.e., polyelectrolytes) or it is associated with other ions present in the solution like H^+. For compounds that are non-ionic, analyte-related ions can be prepared by solution chemistry. Polyethylene glycol, for example, can be analyzed after cationization with alkali ions. The possibility of using water as solvent makes ESI (besides MALDI) a preferred technique for the investigation of biomolecules (proteins, peptides, DNA-sequencing, study of non-covalent complexes).

Trends observable in ESI are the reduction in flow rate and droplet size. Latest achievements in reducing the flow rate to nL/min and the droplet-size also to the nm range, called nanospray [6], allow a characterization at the sub-pmol level. In combination with Fourier transform ion cyclotron (FT-CIR) instrumentation [7] high-resolution spectra with attomoles of analyte can be obtained. Another factor is making the ESI source more robust towards higher flow rates in order to use it as an interface to analytical HPLC. Tolerance of flow rates up to 2 mL/min. has been achieved [8].

A more convenient way is the use of an interface optimized for LC-MS coupling: the atmospheric pressure ionization (API) interface.The use of an ESI-source as a means of on-line detection for field flow fractionation – another separation technique – is described in Ref. [9]. The authors report the analysis of PEGs (M_w 1000–4000 g/mol) and of polystyrene sulfonates (M_w 1400–49,000 g/mol).

2.1.3 Ionization from the Condensed Phase

Synthetic polymers are often non-volatile, cannot be brought thermally to the gas phase without decomposition, and under EI conditons show considerable fragmentation. Thus a direct ionization from the condensed state would be of great advantage. As indicated in Table 2.1, besides electron bombardment (EI) and spray techniques (ESI) ions can be generated by the application of electric fields, by photons and the bombardment with rapidly moving particles.

2.1.3.1 Field Desorption (FD)

Field ionization was first observed in 1953 by Erwin E. Müller at Pennsylvania State University and put into practice as the variation field desorption in 1959 by H.D. Beckey at the University of Bonn [10]. In field desorption the analyte is first deposited onto an activated electrode (emitter) by dipping the electrode into a dilute solution of the analyte or by applying the solution with a syringe. After evaporation of the solvent the emitter is placed into the ion source. The intense electric fields necessary for ionization (up to 10^8 V/cm) are produced by the use of wire-based emitters carrying carbon dendrites of small radius of curvature r (10^3 Å). The field strength E is proportional to V/r. The presence of the intense electric field distorts the potential energy for the interaction of an adsorbed molecule with the emitter surface in such a way that the probability of electron tunneling from the molecule into the emitter electrode is increased dramatically. So M^+ or $[M+H]^+$ ions can be formed at the tips of the emitter microneedles. The energy required for the desorption of an ion under FD conditions is considerably lower than the evaporation energy for neutral molecules. If the sample contains salts, cationization can occur through the attachment of a cation giving $[M+X]^+$. If the sample is vaporized before the area where the field is applied, so-called field-ionization (FI) occurs.

As the sample (and especially emitter) preparation for FD is tedious it has been superseeded for polar and ionic compounds by other techniques like fast atom bombardment (FAB) or MALDI MS. Nevertheless there are some areas of larger non-polar analytes like polyolefins where only FD-MS provides useful spectra [11]. Recent examples are the investigation of epoxy resins containing pyrene moieties [12] and of cyclic mesityl-acetylene oligomers [13].

2.1.3.2 Fast Particles (Fast Atom Bombardment, FAB; Secondary Ion Mass Spectrometry, SIMS)

In the early 1980s [14] a method for the investigation of non-volatile samples, especially for large polar, ionic and/or thermally labile molecules like peptides, was developed: fast atom bombardment, FAB. In this method a beam of noble gas ions (usually Xe or

Ar) produced in an EI-source is accelerated to 6–8 kV and directed through a chamber containing the same atoms. The occuring charge exchange produces a beam of fast atoms as most of the original momentum is maintained in the former ions and neutrals after collision. The process may be simplified as

$$Ar^+_{fast} + Ar_{thermal} \rightarrow Ar_{fast} + Ar^+_{thermal} \qquad (2.9)$$

The thermal ions are deflected after the collision. Of paramount importance was the finding that the best spectra could be obtained when the sample was dissolved in a liquid matrix, e.g., glycerol. This allowed the diffusion of ever new analyte molecules to the surface [15].

Nowadays many of the tasks performed in the past by FAB are performed by MALDI instrumentation. The reason is the fact that the same class of molecules can be tackled, combined with an even higher mass range, a much higher throughput due to automated sample handling, and a broader range of possibilities for synthetic polymers. Also, proteins are investigated much more favorably by MALDI MS.

If, instead of atoms, charged particles are used for sample bombardment, secondary ions characteristic for the sample under investigation are created through a collision cascade as indicated in Fig. 2.4. The technique first developed in the RCA laboratories,

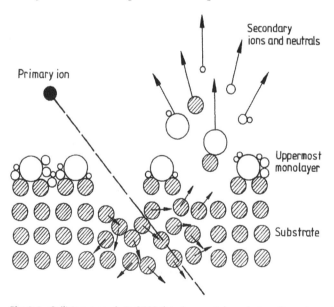

Fig. 2.4. Collision cascade in SIMS showing particle emission after primary ion excitation of keV energies. (Reprinted from [16] with permission of American Chemical Society, U.S.A.)

Princeton, NJ, in the 1950s is called Secondary Ion Mass Spectrometry (SIMS).

Usually Ar-, Cs-, In-, Ga- or O-ions are used as primary particles in pulsed ion guns. Though the principle is similar to FAB, SIMS offers completely different possibilities. As the secondary ion formation is restricted to the surface of the sample, SIMS can be used favorably [16] for surface characterization of bulk materials. In the dynamic mode in which an additional sputter ion source is used, SIMS is frequently used in the electronics industry for the determination of dopant profiles in semiconductor structures. In the so-called static SIMS mode, a technique strongly driven since the 1970s by Benninghoven at the University of Münster, the primary ion beam current is so low that only a small fraction (<1%) of the uppermost monolayer is consumed during the typical recording time of a spectrum (100–1000s). The static mode creates molecular fragments and in favorable cases quasi-molecular ions, thus allowing a molecular characterization of the sample. Phenomena like adhesion, adsorption, corrosion, and bio-compatibility can be investigated. Segregation of additives like antioxidants, slipping agents, and UV-stabilizers in bulk polymers can be studied. Pretreatment procedures to enhance adhesion on technical polymers like corona or plasma treatment can be followed; see Fig. 2.5.

For some classes of polymers like silicones, polystyrene, and perfluorinated compounds, information on the molar mass distribution and possible endgroups in the range up to 10,000 amu can be obtained.

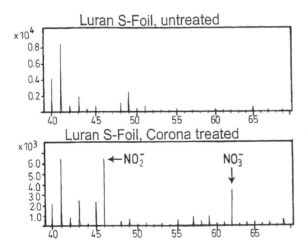

Fig. 2.5. Negative - ion SIMS spectrum of a corona treated sample of Luran (styrene-acrylonitrile copolymer). Characteristic is the occurrence of nitro groups.

Scanning the primary ion beam in a definite way opens up the possibility to obtain laterally resolved atomic as well as molecular images of a surface which could be of interest for semiconductors (dopant distribution), polymeric materials (lateral distribution of additives, migration over interfaces), and for example in biology (distribution of live regulating elements like K, Ca, or drug molecules in biological tissue).

Though there is a variety of applications well documented in the proceedings of the regular SIMS conferences, SIMS is not really widely used in the analysis (molar mass, MMD, endgroups) of synthetic polymers apart from some special applications [17]. A possible reason might be the fact that for many polymers only the repetition units or low molar mass oligomers can be obtained. Although it is called a soft-ionization technique SIMS often leads to considerable fragmentation of the molecules. Last but not least, due to the necessary ultra-high vacuum conditions and the ion source(s), the technique is more expensive than most of the other MS techniques (apart from FT-MS). Recent developments using higher molar mass primary ions like SF_5^+ or Au-clusters [18] cause less fragmentation and enhance the yield in the higher mass range.

2.1.3.3 Photons

Laser radiation can be used for the desorption and/or ionization of analyte molecules. Early attempts were made in the 1970s, for example by M.A. Posthumus et al. at the FOM Institute for Atomic and Molecular Physics (Amsterdam, The Netherlands). Laser desorption (LD) found commercial application in the LAMMA (laser micro mass analyzer) instrument by Leybold-Heraeus, Germany. This instrument was mainly used for the elemental analysis of biological tissue (Ca-, K-, Na-distribution). For reproducibility reasons desorption and ionization are separated. This allows a special post-ionization of the more abundant neutral desorbed species since a typical ion-to-neutral ratio in laser desorption of organic material is about 10^{-3} [19]. Furthermore, a selective ionization with a second laser based on resonance enhanced multi-photon ionization (REMPI) can be used [20]. As attempts to post-ionize laser desorbed species with molar masses greater than 500 amu are not possible without excessive fragmentation, LD methods have not gained importance for synthetic polymers.

In conclusion, the preceeding sections demonstrate that on one hand a variety of ion sources is available, on the other it became evident that there is no universal ionization principle. For a proper use of mass spectrometry for the analysis of synthetic polymers the ionization technique has to be chosen according to the sample to be investigated and the information to be obtained.

2.2 Mass Analyzers

The next logical step following the formation of ions is the separation according to their m/z-value. As for the ion sources several possibilities have been developed for this task. In this section only principles are described that are important in the polymer field with an emphasis on time-of-flight mass analyzers.

Analyzers – according to Nier [21] the heart of a mass spectrometer – are high vacuum areas where the ions extracted from the sample are separated by static or oscillating electromagnetic fields. Ions of different mass, velocity, and/or charge can be collected separately through control of these fields. By the manner in which these are used to achieve ion separation, various types of mass spectrometry can be differentiated. At the beginning of the last century F.W. Aston, Cambridge, and D.W. Dempster, Chicago, obtained important results in atomic and molecular physics through the use of mass analysis (abundance and mass of isotopes and their stability). Dempster developed a magnetic deflection instrument which focussed ions onto a collector, Aston used electric and magnetic fields to focus ions on a photographic plate. These instruments allowed a focusing of the ions with respect to direction and velocity. Due to the achievable high mass resolution, the identification of various organic compounds became possible. As there are numerous descriptions in the literature and the importance of sector field instruments for this book is limited, we will not go into details and describe the following items for ion separation: the quadrupole (Q), the Fourier Transform (FT), and the time-of-flight (TOF) mass analyzers.

2.2.1 Quadrupole Analyzer

Quadrupoles, developed in the 1950s by Paul [22], are the most widespread mass analyzers. Reasons for this are the generally simple set up, lower costs than for other analyzers, and fast scanning compared to sector instruments, which is important for gas chromatographic (GC) detection. Furthemore quadrupoles are compact devices. They differ from other devices as the mass-resolving capabilities result from the ions' intrinsic stability or instability within the analyzer. This is in contrast to the other mass separators, which resolve the ions by dispersing them in space like in magnetic sector instruments or in time as is the case in TOF instruments. Quadrupoles resemble a tunable bandpass filter [23]. Only ions within a narrow mass region, generally below 1 amu, are transmitted through the device. By tuning the position of the bandpass the quadrupole becomes a device for recording mass spectra.

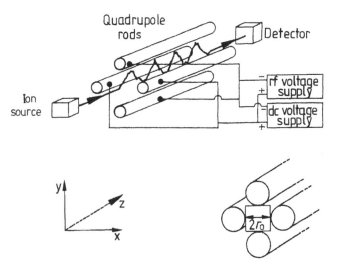

Fig. 2.6. Scheme of a quadrupole analyzer. (Reprinted from [1] with permission of University Press Cambridge, UK)

The quadrupole physically consists of a set of four electrodes, in general circular rods separated by a distance of 2r; see Fig. 2.6. For the filtering of the ions, i.e.; the scanning of the bandpass, a time-independent (the d.c.-voltage U) and a time-dependent (the r.f.-voltage $V\cos(\omega t)$; V=maximum amplitude, ω=angular frequency, t=time) potential are applied. Opposite rods are electrically connected. For an ion to travel from the ion source to the detector through this arrangement it must remain stable (i.e., remain in the space spanned up by the rods) in the two planes given by the two connected electrodes. Under the influence of the combined electric fields the ions follow complex trajectories [24]. To be transmitted to the detector the oscillation of the ions must have finite amplitudes, i.e., the ions may not collide with the rods or with the walls of the analyzer chamber.

By using the parameters $a = 4eU/(\omega^2 r^2 u)$, $q = 2eV/(\omega^2 r^2 m)$, and $\xi = t/2$ the equation of motion for a charged particle entering the quadrupole leads to Mathieu's differential equation

$$d^2u/d\xi^2 + [a_u + 2q_u \cos 2\xi] u = 0 \qquad (2.10)$$

which was derived to describe the vibrational modes of a stretched membrane, where u are the Cartesian coordinates x or y perpendicular to the direction z of the rods. Solutions of this equation may be classified as either bounded or unbounded, the bounded solution corresponding to the case where the displacement of the ions remains stable in both electrode planes x and y.

2.2 Mass Analyzers

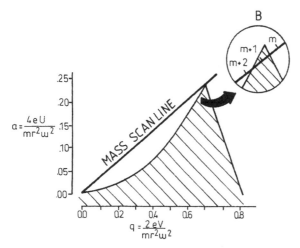

Fig. 2.7. Stability diagram of the quadrupole analyzer. (Reprinted from [23] with permission of Chemical Education Inc., U.S.A.)

The stability only depends on the parameters a and q, so that an a/q-diagram –the stability diagram of the quadrupole – is used to characterize the mode of operation of the quadrupole; see Fig. 2.7. The shaded area below the curve describes the stable a and q values.

Quadrupoles are usually operated in such a manner that the parameters a and q are related by a simple ratio. This condition can be realized by choosing the d.c. voltage to be a certain fraction of the a.c. voltage. Holding the ratio U/V constant means a restriction of the mass filter to operating points which lie on a straight line with zero intercept. This line, the so-called mass scan line, can also be parametrized by mass numbers (m is the only variable for fixed U and V), meaning for the case indicated in Fig. 2.7B, that for ions with mass m+1 the oscillatory movement is stable whereas for m+2 falling into the unstable region of the a/q-diagram the ion is lost in the analyzer assembly. Hence a mass separation is achieved.

The mass spectrum can be scanned by varying U and V but keeping the ratio U/V constant. Another kind of operation varies the frequency of the a.c. voltage while keeping U and V constant. The slope of the scan line can be adjusted by choosing an appropriate U/V ratio, so that the scan line is just intersecting the tip of the stability diagram. In this case only one mass would be able to pass through onto the detector. By lowering the U/V value the mass resolution is reduced. In the limit of the d.c. voltage equal to zero (then a is also zero) one ends up with a line parallel to the q-axis. In this case all ions with an m/z ratio above a certain value have stable trajectories and are transmitted: the quadrupole acts as a high-pass filter. This feature is used with great advantage to provide a collision chamber of high transmission in hyphenated MS systems. Limiting factors for quadrupoles are mass resolution and mass range which nowadays extends up to 6000 amu with typically single mass resolution.

Another device based on the principle of a quadrupole is the ion trap. It allows one to store ions in an evacuated cavity by applying appropriate electric fields to specially formed electrodes (giving a cylindrical cavity) and to eject the ions selectively according to their mass. In ion traps by adding a collision gas an enhanced rate of fragmentation (collision induced dissociation, CID) can be achieved and MS/MS-experiments can be performed. This means that fragmentation of a pre-selected ion left in the chamber (the mother ion) can be performed and the mass spectrum of the fragmentation products be recorded.

2.2.2 Fourier Transform Analyzer

Fourier transform (FT) analyzers exhibit a mass resolution and accuracy higher than any other mass analyzer. It derives from ion cyclotron resonance (ICR) spectroscopy which was incorporated into mass spectrometers in the 1950s [25]. 1979 Marshall et al. [26] adapted Fourier transform methods to ICR spectroscopy and build the first FTMS instruments. In FT instruments the ions are trapped by crossed electric and magnetic fields in a high vacuum chamber; see Fig. 2.8. The analyzer cell is located in the solenoid

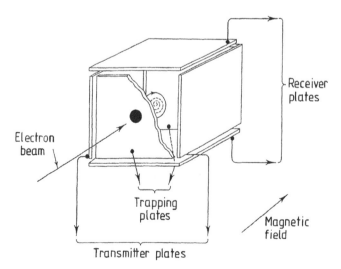

Fig. 2.8. Schematic representation of a cubic FT ion cyclotron resonance analyzer. (Reprinted from [1] with permission of University Press Cambridge, UK)

of a superconducting magnet to achieve a high field strength as mass resolution increases with the field strength.

Magnets between 3 and 9.4 Tesla are in use. Ions in the cubic cell – cylindrically-open devices are also in use [27] – are constrained to move on a circular path perpendicular to the magnetic field axis. They are prevented from escaping axially by the application of a small a.c. field across the trapping plates. The ions are excited to larger orbits by applying an r.f. pulse to the transmitter plates. This ion motion induces image currents in the receiver plates with the ion cyclotron frequency given by $f = zeB/(2\pi m)$ and $\omega = 2\pi f$, i.e., f is proportional to the mass-to-charge ratio of the ion. Accordingly, the ion cyclotron frequency is independent of the ion velocity and thus independent of its kinetic energy. The ion experiences a force that is perpendicular to the magnetic field axis and the direction of its velocity vector, the Lorentz force.

The ionization, mass separation, and detection occur in the same space (the FT analyzer cell) but separated in time. Ions can be created in the cavity itself but also in an external ion source like an ESI or MALDI source.

A common method to detect many masses simultaneously is broadband excitation which is achieved by a rapid frequency sweep, an r.f. chirp (for example by a frequency synthesizer in 1 msec from 10 kHz to 1 MHz). This will excite all ions with ion cyclotron frequencies in the corresponding range. The image current that results from the ions of different m/z values is a composite of sinusodial waves of different frequencies and amplitudes. By performing a Fourier transformation to the time domain the transient signal gives the mass spectrum via the cyclotron equation.

The outstanding feature of FTMS is the achievable mass resolution. Resolution is directly proportional to the duration of the transient signal. Under high vacuum conditions transients of up to 60 s have been observed corresponding to mass resolution in excess of 10^6 in the mass range up to 2000 amu. The mass spectrum of insulin has been recorded with a resolution of 800,000 [28]. In MALDI/FTMS it has been possible to determine the elemental composition of endgroups [29].

The upper mass limit depends on various parameters like the applied magnetic field, the trapping voltage, and length of the cell. Calculations with typical data reveal that ions with masses of up to 950,000 amu could be trapped. In practice, however, much lower values have been achieved due to the fact that resolution is degraded with the smaller ion populations usually encountered at higher masses.

The coupling of FT analyzers to ESI sources might be of great interest as the m/z range of the ions produced (500–2,000 amu) falls in the range where FTMS instruments provide particularly high performance. However, one has to keep in mind that FTMS instrumentation is comparatively complex and expensive, factors which surely have hampered a wide-spread use of this technique especially in the polymer domain.

2.2.3 Time-of-Flight Analyzer

The time-of-flight (TOF) analyzer is the one most frequently used for pulsed ion sources especially in MALDI mass spectrometry. Therefore this device will be described in more detail than the other mass analyzers.

The concept of TOF MS was proposed in 1946 by W.E. Stephens, University of Pennsylvania. The first instruments were commercialized in the 1950s by Bendix Corp., Detroit. The TOF-analyzer makes use of the fact that the mass-to-charge ratio of an ion can be determined by measuring its velocity after accelerating it to a defined kinetic energy; see Fig. 2.9.

The basic equation results from the equality of kinetic and electrostatic energies

$$mv^2/2 = zeU \qquad (2.11)$$

or

$$v = (2zeU/m)^{1/2} \qquad (2.12)$$

with m=mass, z=number of charges, v=ion velocity, U=acceleration voltage. Accordingly, the velocity of an ion with z

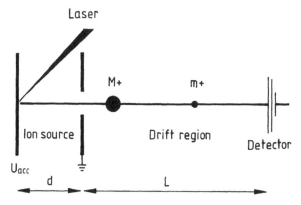

Fig. 2.9. Scheme of the TOF-analyzer. (Reprinted from Biflex II Manual, Bruker, Bremen, Germany)

charges is inversely proportional to the square-root of its mass for a given acceleration voltage U. Thus at a fixed kinetic energy smaller ions travel at a higher speed than larger ones and consequently after passing the field-free drift region L reach the detector earlier. The accelerating sphere d is of the order of 0.5 cm, whereas typical drift lengths range from 50 cm to some meters. The acceleration voltages range from 3 kV to 30 kV so that flight times in the microsecond range result.

Because the ions spend most of their time in the drift region L, their time of flight measured at the detector is, neglecting the time in the ion source

$$t = (m/2eUz)^{1/2} L \qquad (2.13)$$

The time-of-flight spectrum in principle can be converted directly into a mass spectrum according to

$$m/z = 2eU(t/L)^2 \qquad (2.14)$$

As the measured flight time might be different from the actual one due to internal delays in the electronics and other uncertainties, in practice the flight times of at least two ions of known mass (for example H^+ and Na^+ at 1 and 23 amu, respectively) are measured and the spectrum calibrated using the empirical relationship

$$m/z = at^2 + b \qquad (2.15)$$

providing the mass axis for the data acquisition system. The mass resolution in a time-of-flight system is given by [30]

$$\Delta m/m = 2\Delta t/t \qquad (2.16)$$

In contrast to the other mass analyzers which achieve mass separation along a spatial axis, TOF-analyzers achieve m/z dispersion in the time domain. For simple linear instruments mass resolution is of the order 300–400. This rather low value in the past limited the use of TOF instruments. Mass resolving power is limited by small variations in the flight times measured for ions of the same mass (typically on the 100 nsec time scale). A visualization is given in Fig. 2.10 following Cotter [30].

The differences in flight time are mainly caused by distributions in the time of formation of the ions in the source, in the position of ion formation, and in an initial energy spread. A further reason could be the non-ideality of the acceleration fields due to the presence of electrodes and so on. Distributions of ionization

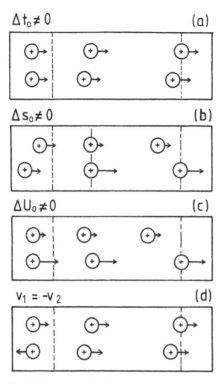

Fig. 2.10. Scheme for the explanation of variations in flight time: **a** two ions formed at different times; **b** two ions formed at different locations; **c** two ions formed with different kinetic energy; **d** two ions with opposite velocities. (Reprinted from [30] with permission of American Chemical Society, U.S.A.)

and detector response time are basically invariant under a given set of experimental conditions (i.e., laser pulse length in the case of MALDI MS) and their relative influence decreases with increasing flight time of the ions, i.e., at higher masses, longer drift tube length L, and lower acceleration voltages.

Due to the sample preparation used in MALDI MS in thin film form on conducting targets the spread in the position of the ion formation is greatly reduced. The remaining factor, initial energy spread, needs some more attention insofar as there are two important experimental means of influencing this factor.

2.2.3.1 The Reflectron

The most effective way of reducing the initial energy spread is by reflecting the ions under a small angle onto the detector; see Fig. 2.11. This procedure, called energy focusing, was suggested by Mamyrin and Shmikk in 1979 [31]. The device, an electrostatic energy mirror, provides one or a cascade of retarding fields be-

2.2 Mass Analyzers

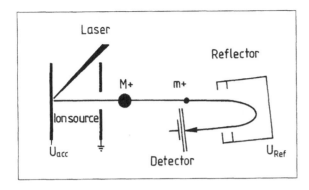

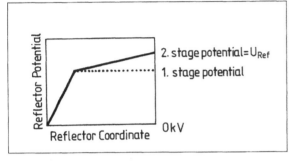

Fig. 2.11. Scheme of the reflectron. (Reprinted from Biflex II Manual, Bruker, Bremen, Germany)

hind the drift region and opposite to the direction of the accelerating field. Higher energetic ions, i.e., faster ions, penetrate more deeply into the opposing field and hence take a longer time to be reflected. Proper alignment of the fields allows to a certain extent a focusing of the ions in space and time. This ion mirror has the additional advantage of increasing the drift length by nearly a factor of two without enlarging the length of the drift tube. Mass resolution is enhanced to several thousand. In the experimental part examples for the use of reflectrons will be given.

2.2.3.2 Time-lag Focusing

Additional reduction in the energy spread of the ions can be achieved by a principle called time-lag focusing, first proposed by Wiley and McLaren in 1955 [32], which has been rediscovered for MALDI applications in the past few years [33]. The idea is to allow a delay between the actual ionization and the acceleration of the ions. In the delay time the velocity dispersion of the ions causes the ion packet to spread. Ions with initial velocity vectors along the axis of the TOF-analyzer will drift to the back or the front of the ion source. When the draw-out pulse is applied, ions with a velocity component in the forward direction are less accelerated then those with opposite velocity components, so that all ions of a given mass should reach the detector at the same time. This correction is dependent on the mass of the ions and enhances the resolving power only over a certain mass range.

2.2.3.3 Practical Consequences

In the reflector mode for masses above 3000 amu the velocity distribution is the resolution limiting factor whereas for smaller masses effects from ionization (moment of ion formation) and detector response time come into play. With 500 MHz digitizers (and faster) it is useful to use the highest possible acceleration voltage (20–35 kV) over the entire mass range to achieve the best results (in some commercial instruments 5 kV acceleration voltage is used).

The measures described from the hardware side allow mass resolution between 10,000 and 20,000 at the mass of insulin (5734 amu) in modern TOF-instruments. This fact – a mass resolution significantly better than with quadropole and near to sector field instruments – together with strong improvements on the data aquisiton side through faster digitizers and computerization make TOF devices the analyzer of choice for pulsed ion sources like in MALDI MS or SIMS.

Unlike the other analyzers (quadrupoles, FT, most sector instruments) in TOF-MS the spectrum is not scanned. On the contrary, the whole spectrum is recorded from one ion packet generated for example by a single laser shot. For the increase in signal-to-noise ratio several spectra are accumulated. Also in contrast to the other analyzers the transmission in TOF devices is very high, reaching more than 50 %; i.e., more than half of the ions entering the drift region will reach the detector. The reasons are evident: no beam defining slits and, because of the temporal separation of the ions, no out-separation of ions in spatial separating devices in front of the detector. With quadrupole instruments at reasonable resolution often only 0.1 % transmission is achieved.

2.3 Detectors

The final step in mass spectrometry after generation and separation of the ions is their detection and processing of the data. As a result of the separation, ion currents with a broad range of intensities from approximately 10^{-9} A to a minimum of about 10^{-18} A, depending on the type of spectrometer and the mode of operation, have to be detected. The devices mainly used for ion detection are the electron multiplier, the Faraday cup, the scintillator together with photomultiplier, and the focal plane detector.

The ions are either directed sequentially to a fixed point in space (Q; TOF) where a single-channel detector (electron multiplier, Faraday cup) is located or dispersed to a plane (sector field instruments) and correspondingly detected by a focal plane detector.

The most common detector, the electron multiplier, can be of discrete or continuous dynode type. The first consists of 10–20 electrodes made from beryllium-copper to achieve a good secondary emission property. The impact of an ion on the first (conversion) electrode releases a shower of electrons which are multiplied in a cascade effect down the dynode structure. Gains between 10^4 and 10^8 can be achieved. In an alternative design the discrete dynodes are replaced by a continuous one which looks like a coiled tapered tube and is called a channeltron electron multiplier. Ions hitting the entrance orifice again release a shower of electrons which are drawn towards the earthed narrow end. Through collisions with the wall of the multiplier the desired cascading effect is achieved.

A very simple effect is used in the Faraday cup: an ion hitting a metal plate connected to earth through a resistor will be neutralized leading to a current flow through the resistor. Conventionally the plate is replaced by a cup in order to capture also the electrons released when the ion strikes the detector. The only disadvantage of this simple and robust detector is its lower sensitivity and response time.

In the Daly detector the electrons released upon ion impact onto an electrode are accelerated after amplification onto a phosphorescent screen, a scintillator, which transfers the ions into photons. These are detected and amplified by a photomultiplier.

Early focal plane detectors were photographic plates. Nowadays often so-called multichannel plates (MCPs) are used. They rely on the fact that the continuous electron multipliers can be miniaturized and arranged side-by-side in array form. They can be used as a spatially resolving detector but often are coupled as in the Daly detector to a scintillator/photomultiplier combination.

As far as detection in TOF-analyzers is concerned, MCP detectors arranged in a stack (dual MCP or Chevron-type) are often used for detection operated at voltages between 1.5 and 1.8 kV. MCP detectors tend to saturate easily, which paraphrases the effect that the charge-up recovery time is longer than the time between two successive ion packets. So the signal response for the second ion cloud is reduced. This is particularly important for the

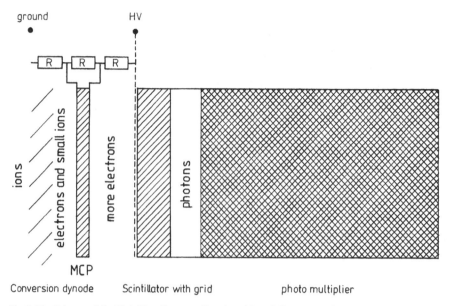

Fig. 2.12. Scheme of the High Mass Detector. (Reprinted from Biflex II Manual, Bruker, Bremen, Germany)

species with higher molar mass of a synthetic polymer distribution as an electron multiplier necessitates a minimum velocity of an incident ion to generate the electron cascade. The ion velocity v is given by

$$v \sim [(U_a - U_z)/(z/m)]^{1/2} \qquad (2.17)$$

where U_a and U_z are the source acceleration voltage and the voltage of the target surface, respectively. Accordingly, the velocity is reduced by the square root of the inverse m/z value. A possible solution to this problem is to increase the velocity of the ions by post acceleration. This is done in MALDI instrumentation by a detector consisting of three items as shown in Fig. 2.12.

The conversion dynode generates fragment ions and electrons from an impinging analyte ion. These smaller charged particles are then accelerated to sufficient energies to cause the electron multiplying effect in the MCP. The grid behind the MCP is held up to 10 kV to assure enough energy for the conversion of electrons into photons in the scintillator.

2.4 Data Processing

Details of the data acquisition and processing depend on the type of mass spectrometer, the mode of operation, the detection system, and the supplier dependent hard- and software, so only some general statements can be given here. In modern systems all key features are computer controlled. After the vacuum systems have reached their setpoints, i.e., after or during sample introduction, all high voltages, detection controls, and so on are under computer control. The detector signals are amplified (in several steps), digitized, and transferred to the main workstation to be saved, displayed, and further processed. The main operations under computer control are the scanning of the mass spectrometer, the data acquisition, and the help in interpretation through the use of mass spectral libraries.

The principle of the TOF analyzer necessitates the measurement of the flight time of the ions, meaning start/stop information is needed. The start signal for pulsed ion sources like in MALDI MS is given by the laser pulse or by the draw-out pulse. The stop signal is given by the arrival of the ions at the detector. There are two principal methods of data acquisition: the first is a fast digitization of the analogue signal produced at the detector (ADC, analogue-to-digital conversion) the second is by single pulse counting.

Regarding integral transient recording, Wiley and Mc Laren [32] recorded their TOF spectra as an oscilloscope trace triggered by the draw-out pulse. The microprobe, introduced in 1975 by Hillenkamp and coworkers [34] for the elemental analysis of biological samples, utilized an 8-bit, 100 MHz waveform recorder (Biomation 8100, Gould, Cupertino, CA) to digitize the spectrum produced from a single laser shot. To improve the S/N ratio a computer was used for the addition of successive analogue mass spectral transients. Integrated transient recording became commercially available on a broader scale through the use of digital oscilloscopes (for example LeCroy, Spring Valley, NY) providing a 1 GHz digitization rate.

Regarding ion counting, ion counting devices are usually based on multi-stop time-to-digital converters (TDC). TDCs detect the onset of pulses (start and stop) and store the time intervals between these events. The time interval measurements from the stop events and the correlated start events are stored in a computer and used to reconstruct a histogram of the number of ions detected in consecutive discrete time intervals. Care must be taken not to saturate the TDC, because if two or more ions coincide in one time channel only one stop event will be detected. Hence TDCs are very suitable for low ion arrival rates.

As stated above, in TOF analyzers one has to take care for the dynamical range of the detection system; see also the remarks made in connection with the MCP detector. Though in principle it is true that a spectrum from a single ionization event can be acquired in about 100 µs one has to perform some averaging over several shots. Typically 100 or 1000 shots are averaged in MALDI MS or SIMS. Rates to record the spectra in MALDI MS are between 10 and 100 Hz given by the repetition rate of the laser system and the flight time of the heaviest ion in the spectrum.

2.5 Coupling of MS Techniques

Interest in the hyphenation of MS techniques evolves from the fact that so-called metastable ions can be detected in the "normal" recording of mass spectra and in particular that these ions can provide additional information for the assignment of unknown species. The term "metastable" characterizes those ions that have just sufficient energy to leave the ion source and fragment before arriving at the detector. It describes the process in which an ion M^+ of mass m_1 yields a fragment A^+ of mass m_2 and a neutral N:

$$M^+ \rightarrow A^+ + N \qquad (2.18)$$
$$m_1/z \quad m_2/z \quad m_1-m_2$$

where m_2 is always smaller than m_1 and the charge has to be conserved. If the fragmentation occurs in the ion source both species are subject to the same acceleration voltage and will be detected as "normal" ion peaks, i.e., there is no means of deciding whether A^+ has resulted from the normal ionization process in the source or from the fragmentation reaction (Eq. 2.18).

If M^+ and A^+ are correlated, one way in achieving the necessary information is to use special scanning techniques in sector field instruments [1,35]. The following points are of principal interest:

- What are typical fragments of a selected precursor ion M^+?
- What are all precursor ions of a selected fragment ion A^+?
- A type of selected reaction monitoring following the reaction $m_1^+ \rightarrow m_2^+$.

A way to create fragments is to collide the ions with a convenient collision partner. This process is called collision induced dissociation, CID, in which part of the kinetic energy of a moving ion is converted to internal energy by collision with background gas molecules. In a second step, fragmentation follows the collisional activation. As these CID processes have a relatively high cross section the amount of fragmented species is higher than in the unimolecular dissociation described above.

The maximum amount of energy available is given in the center-of-mass coordinate system:

$$E_{CM} = E_{LAB}(m_g/(M+m_g)) \qquad (2.19)$$

where M is the mass of the colliding ion, m_g that of the collision partner, and E_{LAB} the ions kinetic energy. From Eq. (2.19) it follows that the transferred energy decreases with increasing mass of the ion colliding and increases with the mass of the collision partner. As internal degrees of freedom for the collision partner have to be avoided (excitation of the collision partner is not the desired process) often noble gases (Ar or Xe) are used as collision gases. Under suitable conditions (gas pressure, collision partner, acceleration voltage) most of the fragment ions will follow the path of the precursor. A simple way to create a collision region is to flood a differentially pumped section of the spectrometer through a small needle inlet from the gas reservoir.

Nowadays many instruments allowing the study of ion fragmentations consist of two independent mass analyzers arranged in "tandem" separated by a collision cell; see Fig. 2.13. This setup is also called MS/MS instrumentation. A particular effective device of this type is the triple quadrupole consisting of two quadrupole analyzers with an r.f.-only quadrupole in between acting as collision cell. Further variations are possible like the electric and magnetic sector of a conventional double focusing spectrometer or, as we will see later, a TOF analyzer behind the collision cell. The use of quadrupoles has the advantage of a velocity independent mass filter, so that the peak broadening upon fragmentation seen in sector field instrument plays no role. The triple quadrupole provides single mass resolution for both the precursor and the fragments.

As MALDI MS is a "soft" ionization technique delivering mainly molecular ions of a sample, MS/MS creates further fragments of a selected species which is very helpful in assigning the various possible endgroups in a polymeric distribution.

With the use of more analyzers, multiple MS variations can be designed. For example three sector instruments can give MS/MS as well as MS/MS/MS spectra (i.e., selected fragment ions are dissociated again) and so on. With FT-ICR and ion traps MS^n experiments can be carried out by using proper driving electronics, i.e.,

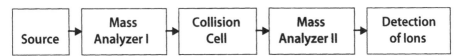

Fig. 2.13. Scheme of tandem mass spectrometry

the fragmentation pathway can be followed through n-1 steps. In these two instruments the steps of the experiments are separated in time and not in space. Precursor ion and neutral loss scans cannot be performed in a simple way in these instruments.

A variety of further hyphenation techniques evolves when the first separation step is not provided by a mass spectrometer but through other mass separating devices, for example gas chromatography (GC), liquid chromatography (SEC or HPLC), field flow fractionation (FFF), and other special types of fractionation techniques. In these setups, which may be used either in an on-line mode but also off-line, MS is used as a detector, broadening the information content of the preceeding separation procedure.

On-line GC/MS systems hold a prime position in analytical chemistry insofar as they allow, besides their high sensitivity and the broad range of applications, the analysis of complex mixtures. In this application the fused silica end of the GC column is fed through a gas-tight heated sheath directly into the ion source of the spectrometer. Other ionization techniques like the spray techniques ESI and APCI greatly enhanced the capabilities of mass spectrometry in LC separations.

2.6 Classical Mass Spectrometry of Polymers

The attempts to use mass spectrometric techniques for the characterization of synthetic polymers date back some decades and of course are not restricted to MALDI MS. Nearly all types of analyzers and ionization principles have been applied to polymers. A comparison of different ionization methods for technical polymers can be found in [36]. An overview on the role of MS in the analysis of polymers is given in [37] and in the book of Lattimer and Montaudo [38]. As has already been pointed out the main obstacle in the case of polymers is the necessity to bring the molecules to the gas phase. Most of the experiments have been performed by using field or laser desorption, pyrolysis, ion bombardment, or spray techniques for this purpose so we will concentrate on these examples.

2.6.1 Field- and Laser-Desorption Mass Spectrometry

Applications of field desorption mass spectrometry (FD-MS) to the direct analysis of synthetic polymers date back to the 1970s. In 1979 Matsuo et al. [39] reported the recording of mass spectra of PS up to 11,000 Da with this technique. Series of PEGs, PPGs, and

2.6 Classical Mass Spectrometry of Polymers

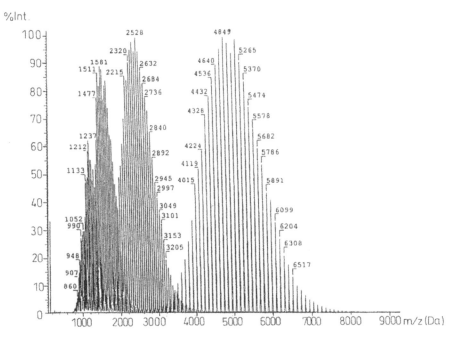

Fig. 2.14. FD mass spectrum of a PS with a molar mass of 5100 g/mol exhibiting singly and multiply charged ions. (Reprinted from [42] with permission of Wiley, UK)

polytetrahydrofurans were examined by Lattimer and Hansen [40]. The fact that hydrocarbon polymers like polybutadiene, polyisoprene, and even PE can be ionized by FD [41] has maintained the interest in this technique. So FD-MS is versatile enough to handle polar and non-polar types of polymers. An example is shown in Fig. 2.14.

As the attainable mass range still remained limited, the method has not achieved a very broad distribution within the polymer community. Some polymer classes prooved to be "poor desorbers" (polyethylene imines or phenol-formaldehyde resins) and the emitter handling remained somewhat tedious.

The same holds true for laser desorption (LD) techniques with respect to polymer analysis. As pulsed lasers are often used, TOF or FT systems are the appropriate types of analyzers [43, 44] The example of an LD-FT mass spectrum of PEG with a molar mass of 6000 g/mol is given in Fig. 2.15. The investigation of several low molar mass polymers (up to 700 Da) like PEG, PPG, PS, and poly(caprolactone) is also described in [45].

A very recent report [46] describes the application of LD-MS to PEs up to 4000 Da. The use of matrices such as retinoic acid or di-

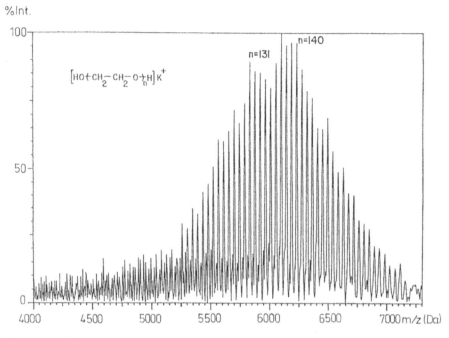

Fig. 2.15. LD-FT mass spectrum of PEG with a molar mass of 6000 g/mol. (Reprinted from [45] with permission of American Chemical Society, U.S.A.)

thranol enhances signal intensity and reduces the amount of fragmentation. The authors claim fragmentation to be the major obstacle to be overcome for the analysis of higher molar mass PEs.

2.6.2 Pyrolysis MS of Polymers

Pyrolysis describes the thermal degradation of (complex) material in vacuum or an inert atmosphere. It causes larger molecules to cleave at their weakest point and thus creates smaller, more volatile fragments, the pyrolysate [47]. Reproducible results can be obtained by heating an appropriate metal substrate used as sample holder to its Curie point. The volatized components of the sample can be ionized, often by low energy electrons or field ionization, and separated according to their m/z values. The technique is called pyrolysis mass spectrometry, PyMS [48], a review on which can be found in [49]. Thermal degradation can also be achieved by simple resistive or laser heating.

Main applications of PyMS are the analysis of complex mixtures and the characterization of the thermal stability of polymers

2.6 Classical Mass Spectrometry of Polymers

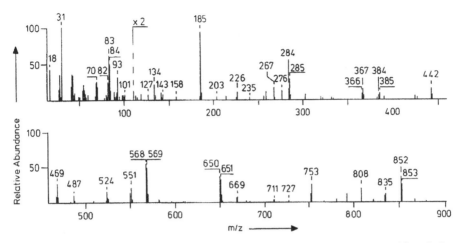

Fig. 2.16. Py-field ionization mass spectrum of poly(hexamethylene sebaccate) (Reprinted from [51] with permission of American Chemical Society, U.S.A.)

Table 2.2. Assigment of the mass signals observed in Fig. 2.16 (n = number of monomeric units, X = dicarboxylic sub-unit, here $(CH_2)_8$)

Mass number	n	Structure of the ion
285	0	[oligomers +H]$^+$
569	1	
853	2	
185	0	[HO(M$_n$)COXCO]$^+$
469	1	
753	2	
267	0	[CH$_2$=CH-(CH$_2$)$_4$-O(M$_n$)COXCO]$^+$
551	1	
835	2	
303	0	[HOCH$_2$CH$_2$-(CH$_2$)$_4$-O(M$_n$)COXCOOH + H]$^+$
587	1	

[50,51]. The analysis of additives like flame retardants [52], hindered amine light stabilizers [53], additives in rubber [54], and bulk polymers like packaging products or paints [55] has been reported.

A typical example from the analysis of polyesters [51] is shown in Fig. 2.16 with the corresponding assignment of the chemical strcutures given in Table 2.2. The repeating unit M with mass 284 Da is O-(CH$_2$)$_6$-OCO-(CH$_2$)$_8$-CO. Abundant series of oligomers appear as [M$_n$]$^{+\bullet}$ and [M$_n$+H]$^+$ ions. Fragmentation leads to carboxonium endgroups. Thermal *cis*-elimination forming one olefinic and one carboxylic endgroup is favored. Altogether a fundamental interpretation of the pyrolytic behavior of polyesters could be given.

Further applications like characterization of biopolymers, amino acids, complex inorganic substances, micro-organisms such as bacteria and fungi, parts of plants, ceramics, and so on are mentioned in [1].

2.6.3 Secondary Ion Mass Spectrometry

Though the main applications of SIMS nowadays can be found in the semiconductor area a variety of attempts are described for polymers. An overview is given by Benninghoven et al. in [16]. Some special classes of polymers like polysiloxanes, perfluorinated compounds, and to a certain extent PS give intense SIMS spectra up to several thousand Da; see for applications [56]. An example for a perfluorinated polyether used as lubricant for data storage media (Krytox) is given in Fig. 2.17.

As SIMS does not necessitate a special sample preparation, the spectrum is directly recorded from the tape surface. Single oligomer resolution extends up to 10,000 Da. Influences of tempera-

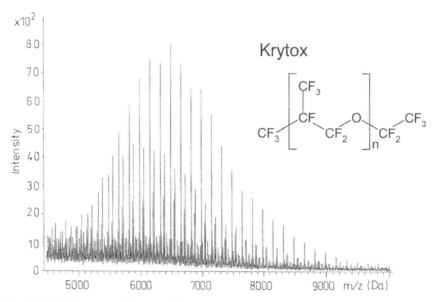

Fig. 2.17. SIMS spectrum of Krytox exhibiting oligomers up to 10,000 Da

ture and sample aging on the oligomer distribution and the chemical composition can be followed directly.

Polyurethanes also have been investigated by SIMS. An example taken from [56] is shown in Fig. 2.18. Four types of peak series labeled A to D can be recognized. The samples have been dissolved in dimethylformamide and spread on a silver target. Therefore and due to a small sodium contamination, cationization through Ag and Na is observed. The origin of the different peak series can be understood from the scheme given in Fig. 2.18. The well-defined fragmentation pattern allows the identification of the structures of the diols, esters, and isocyanates used in the synthesis as well as of the endgroups.

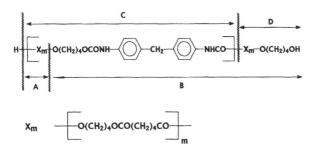

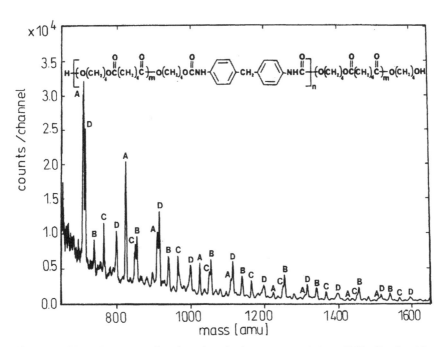

Fig. 2.18. TOF-SIMS spectrum of a polyurethane in the mass range 650 to 1650 Da. (Reprinted from [56] with permission of American Chemical Society, U.S.A.)

In a recent paper it is reported [57] that for poly(styrene-*co*-isoprene) diblock copolymers an improved surface sensitivity (even smaller than 5 Å) by using higher mass fragments is observed. This is attributed to a lower escape depth of high mass ions compared to lower mass ions.

Reihs et al. [58] report the determination of molar masses at the surface of bulk polymers by analyzing fragments originating from repeat units and endgroups. The intensity ratio of these fragments depends on the polymer chain length as demonstrated for bisphenol-A polycarbonate (oligomers disturb the production process of compact discs) and perfluorinated polyether. Further applications of SIMS include the investigation of bio-degradable materials [59], surface seggregation in blends [60] and the surface functionalization of poly(etherketones) [61]. A comprehensive overview on the use of SIMS in practical surface analysis has been published by Briggs and Seah [62]. A collection of reference spectra can be found in [63].

2.6.4 Electrospray Ionization (ESI) Mass Spectrometry

ESI as an extremely soft ionization technique can produce highly charged pseudo-molecular ions in a mass range between approximately 500 and 2500 Da, and in special cases up to 10,000 Da. It has been mainly proven to be successful in biochemistry [64] but is also increasingly applied to synthetic polymers, the first attempts being performed in the 1990s [65]. An overview on applications to non-biological molecules is given by Saf et al. in [66]; a summary of polymers investigated is presented in Table 2.3.

As ESI in general produces higher charged species (charge states up to 25 can be achieved) the polydispersity of polymeric samples is a basic problem due to the variety of mass peaks produced. Reliability, interpretability of the spectra, and discrimination of higher mass species are enhanced by using a chromatographic preseparation.

Table 2.3. Oligomers and polymers investigated by ESI-MS

Polymer	Reference
Poly(ethylene glycol)	[67]
Poly(propylene glycol)	[68]
Polyester and –acrylates	[69]
Polystyrene	[69]
Melamine resins	[70]
Surfactants	[71]

2.6 Classical Mass Spectrometry of Polymers

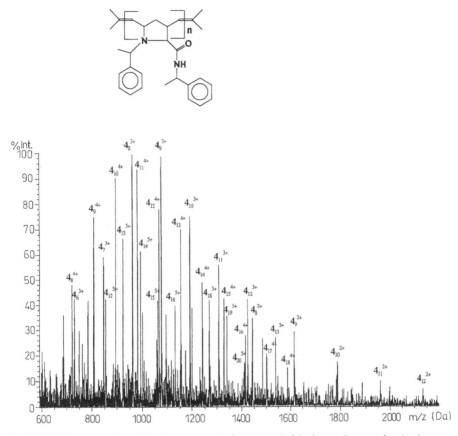

Fig. 2.19. ESI mass spectrum of a polymer sample after removal of the low molar mass fraction by precipitation. (Reprinted from [66] with permission of WILEY-VCH, Germany)

Figure 2.19 shows a typical ESI mass spectrum of the polymer shown above prepared by ring-opening metathesis containing oligomers with degrees of polymerization (n) between 1 and 20. The spectrum shows protonated oligomers $[4 \cdot mH]^{m+}$ written as 4_n^{m+}. After removal of the lower mass oligomers through precipitation, Maziarz et al. [72] report on the investigation of poly(dimethylsiloxanes) by ESI using an FT mass analyzer. The high resolution allows the detailed study of the various possible endgroups of the polymers with (aminopropyl)dimethylsiloxy termini. An example is given in Fig. 2.20. The accurate mass measurements (<10ppm errors compared to theoretical masses) together with a comparison of the experimental and theoretical isotope patterns allows a definite identification of the endgroups; see Table 2.4.

ESI application to acidic solutions of linear and cyclic polyamide-6 oligomers up to octamers are described by Klun et al.

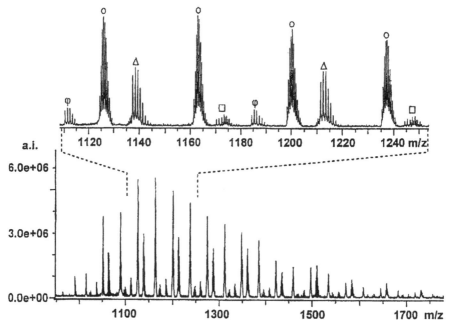

Fig.2.20. ESI-FTMS of poly(dimethylsiloxane) of 2500 Da and proposed molecular structures. (Reprinted from [72] with permission of American Chemical Society, U.S.A.)

Table 2.4. Proposed molecular identities of the oligomer distributions observed in Fig. 2.20

	Suspected oligomer series	n^a	Theor. mass	Exp. mass	Error (ppm)
Δ	$H_3NC_3H_6^+$—[Si(CH$_3$)$_2$—O]$_n$—Si(CH$_3$)$_2$—$C_3H_6NH_2$	13	1137.4067	1137.4121	4.7
O	$H_3NC_3H_6^+$—[Si(CH$_3$)$_2$—O]$_n$—Si(CH$_3$)$_2$—$C_3H_6NH_3^+$	29	2322.7152	2322.7144	0.3
φ	$H_3NC_3H_6^+$—[Si(CH$_3$)$_2$—O]$_n$—CH_3	14	1110.3595	1110.3599	0.4
□	$H_3NC_3H_6^+$—[Si(CH$_3$)$_2$—O]$_n$—H	15	1170.3626	1170.3588	3.2

[a] Degree of polymerization

[73]. High resolution studies using ESI-FTMS for endgroup determiniation of poly(oxyalkylenes) between 400 and 8000 Da are described by Koster et al. [74].

A comparison of the ionization methods MALDI, ESI, and SIMS for poly(dimethylsiloxane) and poly(ethylene glycol) gave, according to Wenyan et al. [36], the result that MALDI exhibits the best performance for reliable molar mass determination and quantification in polymer mixtures.

References
1. JOHNSTONE RAW, ROSE ME (1996) Mass spectrometry for chemists and biochemists. University Press, Cambridge
2. DEMPSTER AJ (1921) Phys Rev 18:415
3. Gaskell SJ (1997) J Mass Spectrom 32:677
4. DOLE M, HINES RL, MACK RC, MOBLEY RC, FERGUSON LD, ALICE MB (1968) J Chem Phys 49:2240
5. FENN JB (1993) J Am Soc Mass Spectrom 4:524
6. WILM MS, MANN M (1994) Int J Mass Spectrom Ion Proc 136:167
7. LOO JA, QUINN JP, RYU SI, HENRY KD, SENKO MW, MCLAFFERTY FW (1992) Proc Natl Acad Sci USA 89:286
8. HOPFGARTNER G, WACHS T, BEAN K, HENION J (1993) Anal Chem 67:2916
9. HASSELOV M, HEELTHE G, LYVEN B, STENHAGEN G (1997) J Liq Chromatogr Rel Technol 20:2843
10. BECKEY HD (1977) Principles of field ionization and field desorption mass spectrometry. Pergamon, Oxford
11. EVANS WJ, DECOSTER DM, GREAVES (1996) J Am Chem Soc Mass Spectrom 7:1070
12. KAJI M, NAKAHARA K, OGAMI K, ENDO T (2000) J Appl Polym Sci 75:528
13. KOJIMA Y, MATSUOKA T, TAKAHASHI H (1999) J Appl Polym Sci 72:1539
14. BARBER M, BORDOLI RD, ELLIOT GJ, SEDGWICK RD, TYLER AN (1981) Nature 293:270
15. DE PANEV E (1986) Mass Spectrom Rev 5:191
16. Benninghoven A, Hagenhoff B, Niehuis E (1993) Anal Chem 65:630A
17. BENNINGHOVEN A, BERTRAND P, MIGEON H-N, WERNER HW (2000) Secondary ion mass spectrometry SIMS XII. Elsevier, Amsterdam
18. KÖTTER F, BENNINGHOVEN A (1998) Appl Surf Sci 133:47
19. GROTEMEYER J, SCHLAG EW (1988) Org Mass Spectrom 23:388
20. BOESL U, NEUSSER HJ, SCHLAG EW (1981) Chem Phys 55:193

21. NIER KA (1998) In: Bud R, Warner DJ (eds) Instruments of science: an historical encyclopedia. Garland, New York
22. PAUL W, REINHARD HP, VON ZAHN U (1958) Z Phys 152:143
23. MILLER PE, DENTON MB (1986) Chem Edu 63:617
24. DAWSON PH (1986) Mass Spectrom Rev 5:1
25. SOMMER H, THOMAS HA, HIPPLE JA (1949) Phys Rev 76:1877
26. MARSHALL AG, COMISAROW MB, PARISOD G (1979) J Chem Phys 71:4434
27. AMSTER IJ (1996) J Mass Spectrom 31:1325
28. MCIVER, RT JR, LI Y, HUNTER RL (1996) Fourth ASMS Conference on Mass Spectrometry and Allied Topics. Portland, OR, 12–16 May
29. VANROIJI GJ, DUURSMA MC, HEEREN RM, BOON JJ, DEKOSTER GG (1996) J Am Soc Mass Spectrom 7:449
30. COTTER RJ (ed) (1994) Time-of-flight mass spectrometry. ACS Symp. Ser. 549
31. MAMYRIN BA, SHMIKK DV (1979) Sov Phys JETP (Engl) 49:762
32. WILEY WC, MCLAREN IH (1955) Rev Sci Instrum 26:1150
33. SCHRIEMER DC, WHITTAL RM, LI L (1997) Marcromolecules 30:1955
34. HILLENKAMP F, KAUFMANN R, NITSCHE R, UNSOLD E (1975) Appl Phys 8:341
35. CHAPMAN JR (1993) Practical organic mass spectrometry: a guide for chemical and biochemical analysis Wiley, Chichester, UK
36. WENYAN Y, AMMON DM, GARDELLA JA, MAZIARZ EP III, HAWKRIDGE AM, GROBE GL III, WOOD TD (1998) Eur Mass Spectrom 4:467
37. KYRANOS JN, VOUROS P (1989) J Appl Polym Sci Appl Polym Symp 43:211
38. LATTIMER RP, MONTAUDO G (2001) Mass spectrometry of polymers. CRC Press, Boca Raton, USA
39. MATSUO T, MATSUDA H, KATAKUSE I (1979) Anal Chem 51:1329
40. LATTIMER RP, HANSEN GE (1981) Macromolecules 14:776
41. LATTIMER RP, SCHULTEN H-R (1983) Int J Mass Spectrom 52:105
42. ROLLINS K, SCRIVENS JH, TAYLOR MJ, MAJOR H (1990) Rapid Commun Mass Spectrom 4:355
43. MATTERN DE, HERCULES DM (1985) Anal Chem 57:2041
44. COTTER RJ, HONOVICH JP, OLTHOFF JK, LATTIMER RP (1986) Macromolecules 19:2996
45. BROWN RS, WEIL DA, WILKINS CL (1986) Macromolecules 19:1255

46. Chen R, Yalcin T, Wallace WE, Guttman CM, Li L (2001) J Am Soc Mass Spectrom 12:1186
47. Irwin WJ (1982) Analytical pyrolysis: a comprehensive guide. Marcel Dekker, New York, USA
48. Meuzelaar HLC, Haverkamp J, Hileman FD (1982) Pyrolysis mass spectrometry of recent and fossil biomaterials. Elsevier, Amsterdam
49. Blazsó M (1997) J Anal Appl Pyrolysis 39:1
50. Kubatovics F, Blazsó M (2000) Macromol Chem Phys 201:349
51. Plage B, Schulten H-R (1990) Macromolecules 23:2642
52. Ebdon JR, Price D, Hunt BJ, Joseph P, Gao F, Milnes GJ, Cunliffe LK (2000) Polym Deg Stab 69:267
53. Blazsó M (2001) J Anal Appl Pyrolysis 58/59:29
54. Lattimer RP, Harris RE, Rhee CK, Schulten H-R (1986) Anal Chem 58:3188
55. Wilcken H, Schulten HR (1996) Anal Chem Acta 336:201
56. Bletsos IV, Hercules DM, Van Leyen D, Benninghoven A (1987) Macromolecules 20:407
57. Mehl JT, Hercules DM (2001) Macromolecules 34:1845
58. Reihs K, Voetz M, Kraft M, Wolany D, Benninghoven A (1997) Fresenius J Anal Chem 358:93
59. Brinen JS, Greenhouse S, Jarrett PK (1991) Surf Interface Anal 17:259
60. Lhoest JB, Bertrand P, Wenig LT, Duvez JL (1995) Macromolecules 28:4631
61. Henneuse-Boxus C, Poleunis C, De Ro A, Adriaensen Y, Bertrand P, Marchaud-Brynaert J (1999) Surf Interface Anal 27:142
62. Briggs D, Seah MP (1992) Practical surface analysis, vol 2. Ion and neutral spectroscopy. John Wiley, Chichester, UK
63. Briggs D, Brown A, Vickerman JC (1989) Handbook of static secondary ion mass spectrometry (SIMS). Wiley, Chichester, UK
64. Hofstadler SA, Bakhtiar R, Smith RD (1996) J Chem Educ 73:A82
65. Nohmi T, Fenn JB (1992) J Am Chem Soc 114:3241
66. Saf R, Mirtl C, Hummel K (1997) Acta Polymer 48:513
67. Saf R, Mirtl C, Hummel K (1994) Tetrahedron Lett 36:6653
68. Smith RD, Cheng X, Bruce JE, Hofstadler SA, Anderson GA (1994) Nature 369:137
69. Nielen MWF (1996) Rapid Commun Mass Spectrom 10:1652
70. Nielen MWF, van de Ven HJFM (1996) Rapid Commun Mass Spectrom 10:74

71. Parees DM, Hanton SD, Willcox DA, Clark PAC (1996) Polym Prepr ACS Div Polym Chem 37:321
72. Maziarz EP III, Baker GA, Wood TD (1999) Macromolecules 32:4411
73. Klun U, Andrensek S, Krzan A (2000) Polymer 42:7095
74. Koster S, Duursma MC, Boon JJ, Heeren RMA (2000) J Am Soc Mass Spectrom 11:536

3 Fundamentals of MALDI-TOF Mass Spectrometry

In Chap. 2 the main mass spectroscopic instrumentation with relevance to the analysis of synthetic polymers has been described, demonstrating that there is no single means solving all problems. MALDI MS now adds another facet to the possibilities for mass spectroscopic characterization of polymers. The main features are the extremely high mass range in combination with a useful mass resolution as compared to all other MS techniques.

3.1 The MALDI Process

Traditionally mass spectroscopic techniques have been of limited use for the investigation of higher mass species apart from fast atom bombardment for proteins. All MS techniques require that the sample under investigation is transferred into isolated, intact ionized species in the gas phase. Since the 1960s laser radiation has been applied for the generation of ions in mass spectrometers [1]. Two main principles evolved through these attempts to achieve laser desorption/ionization (LDI): (a) the best results are obtained when resonant absorption occurs at the laser wavelength, corresponding to exciting electronic states for far UV lasers and exciting vibrational states for IR lasers; (b) to avoid thermal decomposition the energy must be transferred in a short time interval, i.e., lasers with pulse widths between 1 and 100 nsec are used.

Short pulses in combination with the possibility to focus to small spot sizes compared to the other dimensions of the ion source result in ion generation comparable to a point source in space and time. These conditions are ideal for TOF analyzers.

Depending on the molecular structure and laser radiation the generation of ions by laser desorption (LD) remained limited to about 1000 amu for biopolymers as well as synthetic polymers. One reason for this limitation in resonant desorption experiments is the fact that energy is also transferred into dissociative channels. The main breakthrough towards higher masses was achieved by embedding the analyte molecules in low concentration into a solid (Millenkamp et al.) [2,3] or liquid (Tanaka, Nobel Price winner for chemistry in 2002)[4] highly absorbing matrix. In this way an efficient and controllable energy transfer was realized by guarding the molecules from excessive amounts of energy. In this case the matrix serves as a mediator for energy absorption. Application of this MALDI principle led to a drastic extension of the accessible mass range. For proteins a mass range up to 300,000 amu was achieved [5]; for synthetic polymers recent results demonstrated that even polystyrene with masses of 1,500,000 amu can be measured [6].

The progress in MALDI has been driven mainly by empirical advancement, the underlying physico-chemical processes still being under debate. Experimental findings are the following [7]: (1) essentially only singly charged ions are observed; (2) singly charged positive or negative ions are observed depending on the polarity chosen for the extraction field; (3) radical molecular ions are detected for most common UV matrices; (4) for completely different regimes of laser wavelength (UV vs IR) very similar analyte mass spectra are obtained; and (5) the spectra are characterized by a low amount of prompt fragmentation, i.e., MALDI belongs to the so-called soft ionization techniques. The occuring fragments are generated by unimolecular (metastable) decomposition and by collisions with the background gas (post source decay, PSD).

A collection of radiation sources used is given in Table 3.1. The choice of the matrix is crucial for MALDI MS and will be described in Sect. 3.2. It was found that pulsewidths of the laser up to several tens of nsec have little influence on the MALDI mass spectra. This finding indicates that the desorption/ionization process is determined by the laser fluence (J/cm^3) rather than by the irradiance (W/cm^2).

The MALDI process, especially the ionization step itself, is in contrast to the impressive range of applications far less understood, although the desorption/ionization process has been studied in detail [8,9]. It is generally accepted that the laser excitation results in the ablation of a surface layer of the matrix/analyte solid solution. Based on the matrix and analyte ion velocities the formation of a gas jet carrying the analyte is plausible. This MALDI-"plume" expands in a strongly forward directed fashion. Through

Table 3.1. Lasers used for MALDI MS

Type of laser	Wavelength (nm)	Pulse Width (nsec)	Photon Energy kcal/mol	eV
Nitrogen	337	>1	85	3.68
Nd:YAG (4x)	266	5	107	4.66
Excimer (ArF)	193	15	148	6.42
CO_2	10,600	100	2.7	0.12

the use of time-delayed extraction of the ions the direct determination of the initial ion velocity became possible [10,11]. Typical data for proteins – depending on the matrix used – are between 273 m/sec (hydroxyphenylazo benzoic acid, HABA) and 543 m/sec (2,5-dihydroxy benzoic acid, DHB). These findings strongly support the discussion on the initial velocity spread and the consequences in TOF analyzers. One hypothesis resulting from a high matrix velocity with analyte incorporation tells that expansional cooling in the MALDI plume is responsible for the soft desorption process. A review on the present knowledge on the ionization steps has been given by Zenobi and Knochenmuss [12].

3.1.1 Primary Ion Formation

In the positive-ion mode (extraction voltage negative) commonly radical cations, protonated pseudo-molecular ions, and cationized pseudo-molecular ions are formed. A consideration of the energetics of the generation of free ions from neutral molecules in vacuum reveals that this process would be highly endothermic. The energy for the creation of a proton from the matrix (CH or OH bonds) would be 14 eV or 322 kcal/mol. The separation of an Na^+Cl^- ion pair in vacuum would cost 4.8 eV or 110 kcal/mol. In a salt crystal this value has to be multiplied by the Madelung constant of 1.75 thus giving 8.4 eV. These values are much higher than that provided by the nitrogen lasers usually applied in commercial instrumentation (see Table 3.1). Dielectric screening by the matrix, residual solvent, or non-ionic parts of the large analyte molecules dramatically reduce the Coulomb energy. The exact energy cost of ion separation in the highly polar condensed phase MALDI sample is hard to predict but it is certainly several eV less than in the gas phase. Following Israelachvili [13] the ion dipole energy of Na^+ complexed with single water molecules is 1 eV; for RO-H ionic separation a value of 4.3 eV has been found to be consistent with the threshold for proton transfer in small gas phase clusters [14]. In a recent paper [7] direct evidence is given that the main process reducing the ionization energy may be cluster formation (analyte, matrix, counter ions) in accordance with results concerning the photoionization properties of small neutral matrix clusters [15,16].

The most straightforward explanation for laser generated ions would be multiphoton ionization (MPI; absorption of n photons hν) leading to a matrix radical cation:

$$M \xrightarrow{n(h\nu)} M^{+\bullet} + e^- \quad (3.1)$$

The ionization potentials (IP) for matrix molecules in solid crystals or larger matrix aggregates are still not known but from molecular beam techniques and two-color two-photon laser spectroscopy [15] the IP of DHB has been determined to be 8.05 eV, for the dimer 7.93 eV, and for the water adduct 7.78 eV; that is, direct photo-ionization of the matrix is not a favorable process.

A more likely process would be energy pooling which means that two (or more) separately excited matrix species combine their energy to yield one highly excited matrix molecule or a matrix radical cation [12];

$$MM \rightarrow M^*M^* \rightarrow M + M^{+\bullet} + e^- \qquad (3.2)$$
$$2h\nu$$

or

$$MM + A \rightarrow MM + A^{+\bullet} + e^- \qquad (3.3)$$
$$h\nu$$

This is in accordance with experiments of Land and Kinsel [16] for a multicenter mechanism.

The most frequently proposed ionization mechanism dates back to the roots of MALDI [2]: excited-state proton transfer (ESPT). The following steps are discussed:

$$M + h\nu \rightarrow M^* \qquad (3.4)$$

$$M^* + A \rightarrow (M - H)^- + AH^+ \qquad (3.5)$$

$$M^* + M \rightarrow (M - H)^- + MH^+ \qquad (3.6)$$

In many of the effective UV matrices a hydroxyl group is in ortho position to a carbonyl moiety (COOH, CO, $CONH_2$) facilitating proton transfer (intramolecular and to the analyte). The fact that often positive and negative ions can be generated by the same matrix can be explained by assuming a disproportionation reaction:

$$2M \rightarrow (MM)^* \rightarrow (M - H)^- + MH^+ \qquad (3.7)$$
$$n(h\nu)$$

Thermal ionization is unlikely in bulk matrices insofar as the rather high temperatures necessary would not be available under general MALDI conditions [12].

3.1.2 Secondary Ionization

As already described, the UV-MALDI plume can be characterized as a rapid solid-to-gas phase transition [8] with densities of probably several percent of the solid density up to 100 nsec after the laser pulse. So the primary ions are created in a moderately hot bath of neutral matrix molecules and clusters can undergo collisions and consequently secondary ion-molecule reactions leading to protonated or cationized analyte species. These last two steps are the main secondary processes occuring in the gas phase of the MALDI plume.

3.1.2.1 Gas Phase Proton Transfer
In matrix-matrix reactions like

$$M^{+\bullet} + M \rightarrow MH^+ + (M-H)^{\bullet} \qquad (3.8)$$

species may be created which can cause proton transfer onto the analyte:

$$MH^+ + A \rightarrow M + AH^+ \qquad (3.9)$$

According to Harrison [17] the process is characterized by $\Delta G < 0$ and very efficient in the gas phase. As ΔG values are often not available, proton affinities (PA, $\sim \Delta H$) are used instead. The values measured up to now vary between 183 to 225 kcal/mol; see Table 3.2.

Proton affinities of many peptides, proteins and polymers are higher especially if they are carrying basic entities. Aromatic or oxygen-containing analytes may have lower proton affinities than the common matrices, so choosing a matrix with a lower proton affinity will promote ionization.

Table 3.2. Proton affinities of MALDI matrices according to Zenobi and Knochenmuss [12], structures are given in Table 3.3

Compound	Proton affinity measured (kcal/mol)	Gas-phase basicity measured (kcal/mol)
Ferulic acid	196.5	
4-Hydroxy-α-cyano cinnamic acid	202.5	215
2,5-Dihydroxybenzoic acid	204	197
2-(4-Hydroxyphenylazo)-benzoic acid	204	
Dithranol	209	
Sinapinic acid	210	206
trans-3-Indole acrylic acid	215	

As gas phase basicities for typical deprotonated matrices (M–H) are higher than 300 kcal/mol [12] the process

$$(M - H)^- + A \rightarrow M + (A - H)^- \qquad (3.10)$$

is not very likely for nitrogen-containing compounds, alcohols, carbonyls, thiols [17].

3.1.2.2 Gas Phase Cationization

Another way of achieving ionization, especially in the case of synthetic polymers, is the addition of alkali salts to the matrix solution [18]. Cation affinities are in a range of 25 to 40 kcal/mol for smaller molecules, i.e., substantially smaller than the corresponding proton affinities.

MALDI MS of polystyrene gives another hint into gas phase ionization. This polymer is best ionized by the addition of silver or copper salts; see section on sample preparation. It is assumed that the d-orbitals of the transition metals bind preferentially to the phenylic π-system of the polymer. In experiments of Lehmann et al. [19] the cationization agent, a Cu(II) salt, was separated from the analyte/matrix mixture by a layered sample preparation technique. Nevertheless strong signals of Cu-cationized PS were recorded. Protonation on the other hand would be an endothermic process. For the matrix dithranol, which is one of the preferred matrices for PS, and using the proton affinity of benzene as an upper aproximation for PS we have H = PA (dithranol) – PA (benzene) = 209 kcal/mol – 181 kcal/mol = 28 kcal/mol.

In a recent paper Karas et al. [7] gave evidence that the ionization process may be understood as an initial formation of charged clusters between matrix and analyte which relax through ion-ion and/or neutral-ion chemical reactions to singly charged species. This might explain the experimental finding that in MALDI MS mainly singly charged species are observed, as higher charged clusters cannot survive in the gas phase above the MALDI target (neutralization is considered to be the dominant process).

3.2 Matrices and Sample Preparation

Matrices and preparation techniques described in the literature are numerous and retain a slightly alchemistic touch. From the general principles described in Sect. 3.1 some rules of thumb can be derived simplifying the crucial choice of the proper matrix. If there is an idea of the acidity or basicity of the analyte, a matrix with an appropriate proton affinity or gas phase basicity can be selected from the tables.

3.2 Matrices and Sample Preparation

If cationized analytes are to be expected, a matrix should be chosen which does not compete for the cation with the analyte. In the case of aromatic entities present in the analyte, AgTFA or Cu(II)Cl$_2$ are the preferred cationizing agents. A list of typical matrices is given in Table 3.3. As can be inferred from the table most of the matrices contain aromatic moieties assuring a sufficient absorption of the laser radiation (in general 337 nm from a N$_2$ laser). Though a great variety of matrices has been described most of the results on synthetic polymers have been obtained by using dithranol, HABA, DHB, or IAA. A matrix showing little clustering and adduct formation which is well suited for the investigation of low molar mass oligomers is 4-hydroxybenzylidene malonitrile (HBM). It can also be used for PS, PMMA, and polyacrylonitrile [20]. Liu and Schlunegger [21] described various azo compounds (HABA, 4-nitroazobenzene, 4-phenylazoresorcinol) for the detection of synthetic polymers especially for polybutadiene.

Table 3.3. List of frequently used matrices, highly soluble (+), soluble (o), not soluble (−)

Matrix	MW	Name	Structure	Solubility
Dithranol	226.23	1,8,9-Trihydroxy-anthracene		+: HFIP, THF, DMF, xylene, DMA, toluene, Cl-benzene, NMP, CHCl$_3$, 0: acetone, DMSO, hexane, EtOH −: ACN, H$_2$O, HCOOH
HBM	170.05	4-Hydroxy-benzylidene malonitrile		+: THF, DMA, EtOH, HFIP 0: H$_2$O
CCA	189.17	α-Cyano-4-hydroxy cinnamic acid		+: THF, DMF, EtOH, DMA, acetone 0: xylene, toluene −: H$_2$O, HFIP, Cl-benzene
NA	223.23	9-Nitroanthracene		+: toluene, THF, DMF, ACN, DMA CHCl$_3$, DMSO, NMP, Cl-benzene, 0: hexane, HFIP, EtOH, MeOH −: H$_2$O, HCOOH
HABA	242.23	2-(4-Hydroxy-phenylazo) benzoic acid		+: THF, acetone, DMF, NaOH, NMP 0: EtOH, HFIP −: CHCl$_3$, toluene, hexane, H$_2$O, ACN, HCOOH, xylene
DHB	154.13	2,5-Dihydroxy benzoic acid		+: THF, acetone, DMF, EtOH, NaOH, NMP, H$_2$O/ACN 1:1 0: H$_2$O, HFIP, toluene −: CHCl$_3$, Cl-benzene

Table 3.3. List of frequently used matrices, highly soluble (+), soluble (o), not soluble (–)

Matrix	MW	Name	Structure	Solubility
IAA	187.20	3,β-Indolacrylic acid		+: THF, acetone, DMF, MeOH, EtOH, HCOOH –: toluene, hexane, H_2O, ACN, $CHCl_3$, xylene
4NA	138.13	4-Nitroaniline		+: THF, EtOH, DMF, ACN, acetone, HCOOH, HFIP 0: H_2O –: NaOH, $CHCl_3$, toluene
9-ACA	222.24	9-Anthracene carboxylic acid		+: THF –: H_2O, ACN, HFIP, HCOOH
SA	224.21	Sinapinic acid		+: DMF, THF 0: EtOH, HFIP –: toluene, H_2O, ACN
MSA	168.15	5-Methoxy salicylic acid		+: ACN/H_2O 1:1, ACN, DMF 0: EtOH, toluene, HFIP –: hexane, H_2O
CSA	172.57	Chlorosalicylic acid		+: THF, DMF, EtOH, $CHCl_3$, ACN/H_2O 1:1, NMP 0: toluene –: hexane, H_2O
AHB	153.14	4-Amino-3-hydroxy-benzoic acid		+: DMF, EtOH 0: MeOH –: H_2O, toluene
AP	95.11	Aminopyrazine		+: H_2O, EtOH, acetone, THF, HFIP –: toluene
ANP	139.11	2-Amino-5-nitropyridine		+: THF, HFIP 0: H_2O, EtOH –: toluene
AMNP	153.14	2-Amino-4-methyl-5-nitropyridine		+: DMSO/Phenol –: THF, NaOH, H_2O
PA	139.11	3-Hydroxy-picolinic acid		+: HFIP, DMF –: THF, toluene, EtOH, H_2O

The effect of the pH has been pointed out by Dogruel et al. [22]. DHB, for example, functions best as a proton donor at its intrinsic pH whereas salts of the matrix do not allow the recording of spectra. A variety of matrices allowing the preparation of useful MALDI samples in the pH range between 2 and 11 have been described by Fitzgerald et al. [23].

Sample preparation is an important step in MALDI MS especially for synthetic polymers which always have a distribution of different chain lengths (polydispersity). In contrast to proteins, no standard protocols have been developed because of the diversity of polymeric materials available. In general, matrix and sample preparation should serve three aims: (A) co-crystallization with the analyte in a large molar excess to prevent cluster formation of analyte molecules; (B) strong absorption of the exciting laser light in the matrix with subsequent energy transfer to the analyte; and (C) achieving ionization of the analyte.

The most widespread sample preparation procedure is the so-called dried droplet method which is schematically shown in Fig. 3.1. Appropriate amounts of matrix and polymer are dis-

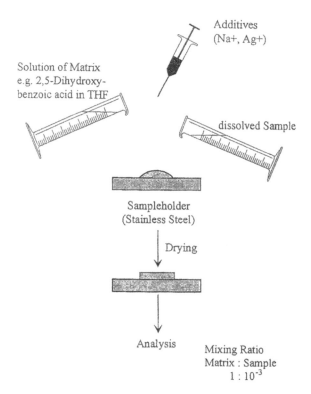

Fig. 3.1. Sample preparation for MALDI MS: dried droplet procedure. (Reprinted from [33] with permission of WILEY-VCH, Germany)

solved in compatible, preferentially identical solvents and mixed to achieve a molar ratio of analyte/matrix between 1:100 to 1:1000. Typically a 20 g/L solution of matrix and a 1 to 10 g/L solution of analyte are prepared, and if necessary a 0.1 molar solution of the cationization agent. The three solutions are mixed in a ratio of 10(matrix):10(analyte):1(cationization agent). Of this solution, 1–5 µL are spotted onto the stainless steel target (some suppliers coat their sample holder by a thin gold layer). The sample is allowed to dry and then introduced into the ion source of the spectrometer. To enhance solvent evaporation the use of a fan is described in the literature.

To improve sample homogeneity further techniques are reported:

1. Spin-coating of the samples [24].
2. Formation of thin films in a layered manner, where first the matrix is deposited, and subsequently the analyte [25,26]. It has been discussed that crushing the pure matrix layer may provide a favorable template for further analyte/matrix growth. It is also reported [24] that crushed spin-dried samples have a lower ionization threshold fluence than the standard dried-droplet preparations.
3. Simple air-spray deposition is described as a method giving excellent shot-to-shot and spot-to-spot reproducibility. The method utilizes an air-spray brush (Azteki Inc., Rockford, USA) and prevents problems with precipitation which might occur at the edges of the sample due to inhomogeneous solvent evaporation. An automated device for spraying has been commercialized [27] and is used for example in the on-line coupling to SEC instrumentation. This principle is described extensively in Chap. 7.
4. Electrospray deposition also provides homogeneous films but is more complicated to use and might cause fragmentation of the analyte [28]. The set-up is similar to the one described on ion sources and spray techniques in Chap. 2. A device combining MALDI and atmospheric pressure ionization is described by Laiko et al. [29]. Sensitivity and stability are enhanced when the ion transfer from atmospheric pressure to high vacuum is assisted pneumatically by a stream of nitrogen. The same matrices and analyte/matrix ratios as in conventional MALDI MS can be used. It is claimed that the technique provides an even softer ionization than normal vacuum MALDI.
5. As solvent evaporation might cause analyte fractionation and not all polymers are soluble in suitable liquids there are ongoing efforts to achieve simplified sample preparations for

MALDI MS. Skelton et al. [30] describe the simple mixing of finely ground powders of analyte and matrix. Their solid sample preparation is similar to making a crystalline KBr pellet for infrared analysis. Empirically it was found that the mixing ratio of analyte to matrix should be around 10:1 to 1:1. Results are described for the analysis of polyamides (1000 to 4000 amu) using 3-aminoquinoline as matrix. Further attempts with these so-called pellet preparations are performed with the aim to elucidate the question or wether the analyte molecules are expelled from the matrix crystals [31]. Measurements of the initial velocities point to different ionization/desorption processes for different classes of analytes [11].

A totally different way of sample preparation is described by Siuzdak et al. [32]. The method is called **Desorption/Ionization on Porous Silicon Mass Spectrometry (DIOS-MS)**. A porous silicon surface, achieved by the selection of special silicon wafers and an electrochemical etching step, is used as matrix substitute: the analyte solution is directly spread onto the silicon. So spectra without matrix peaks have been obtained for a variety of biomolecules. Preliminary experiments revealed that PEG (see also [32]) and PS of molar masses of some thousand amu can be investigated. DIOS substrates became commercially available recently. For the field of synthetic polymers in general the accessible mass range seemed to be a little low.

The preceeding listing demonstrates that the preparation of MALDI samples has been – and still is – a point of debate and research. Nevertheless most of the results described in the literature and performed in the labs of the authors of this book have been obtained by the dried-droplet procedure.

3.3 Interpretation of Spectra

Prerequisite of specta interpretation is the mass calibration of the MALDI instrument in order to make use of one of the advantages, the absolute mass scale. In the beginning of MALDI MS application often internal standards (e.g., protein mixtures) during each or after a certain number of measuring cycles were recorded (one of the sample spots was then devoted to calibration). With the development of higher resolution instruments and more stable voltage sources the following procedure is generally accepted. As described in the introduction on TOF-analyzers the basic equation connecting mass-to-charge ratio m/z with the flight time t is $m/z = at^2 + b$ with a and b being constants, which can be determined using at least two ions of known mass/charge ratio for calibration. Low molar mass ions are for example the alkali ions Na^+ or K^+ (23 or 39 amu), typical matrix ions or clusters thereof in the range of 100 to several 100 amu and insulin (5734 amu) in the high mass range. The calibration can be refined if a polymer is used (e.g., PS or PMMA) with known maximum of the molar mass distribution falling into a mass range with single oligomer resolution. An example is described below; see Fig. 3.3.

In the mass range where single polymer chains can be resolved mass spectrometry can provide the following information: mass of the constituent repeat units, endgroups, e.g., chemical heterogeneity (molar mass of by-products), and the molar mass average M_n.

An example for the analysis of an anionically polymerized PS is given in Figs. 3.2 and 3.3. The enlarged MALDI-TOF mass spectrum is shown in Fig. 3.3, displaying a molar mass distribution with a maximum, the most probable peak M_p, around 8800 amu. The peak-to-peak distance taken from this enlarged spectrum amounts to 104.15 amu and reflects the PS repeat unit.

Since cationization occurred through the attachment of silver ions the mass of silver, 108 amu (isotopes 107 and 109), has to be substracted from the mass number in the spectrum giving for the most probable peak 8598 amu. Divided by the mass of the repetition unit 104.15 amu gives a value of 82 for the number of PS units of this oligomer, leaving a remainder of 58 amu. This residual mass can be assigned to the mass of the endgroups. With the knowledge of the polymerization mechanism a butyl group and a hydrogen atom are plausible endgroups. M_n and M_w values can be calculated according to the formulae given in Eqs. (1.1) and (1.2).

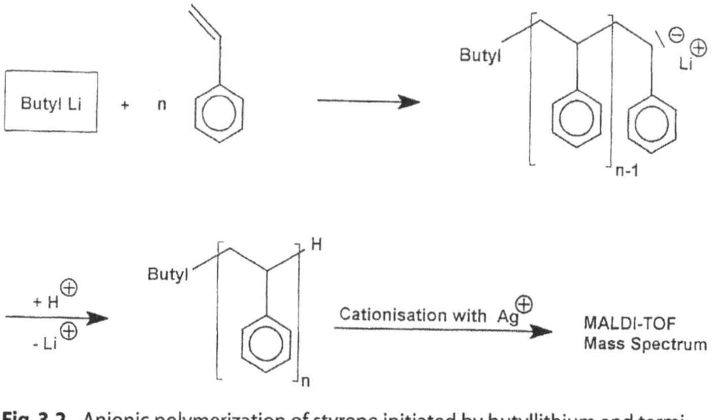

Fig. 3.2. Anionic polymerization of styrene initiated by butyllithium and terminated by quenching in methanol, ionization is achieved by adding a silver salt. (Reprinted from [33] with permission of WILEY-VCH, Germany)

3.3 Interpretation of Spectra

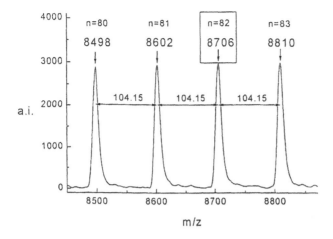

$\boxed{8706} = 82\,(n) * 104.15\,(\text{Styrene}) + 57\,(\text{Butyl})$

$1\,(\text{Hydrogen}) + 108\,(\text{Ag}^+)$

Fig. 3.3. Enlarged section of the MALDI spectrum of an anionically polymerized PS with a molar mass of 9000 g/mol; matrix: dithranol; cationizing agent: silvertrifluoroacetate; solvent: THF. The *bottom part* exemplifies the calculation of the possible endgroups. (Reprinted from [33] with permission of WILEY-VCH, Germany)

Often computer software of the suppliers is available to calculate these data. Obviously information on endgroups and repeat units is lost if the molar mass of the sample is so high that resolution of single oligomers is not achieved (for PS resolution of single oligomers up to 50,000 amu is possible) or if the number of endgroups and/or different oligomers in statistical copolymers is so high that a peak at (nearly) every mass results.

Information on the molar mass averages is falsified if the distribution of oligomers is not recorded correctly. In general this is the case if the polydispersity is too high. Depending on the mass range already polydispersities > 1.3 can be problematic. This point and possible remedies will be described in the next section where the influence of external parameters on the spectra will be discussed.

3.4 Experimental Parameters

This section reviews the main parameters influencing the appearance of MALDI mass spectra. The importance of a proper matrix choice has been pointed out before; see Sect. 3.1. The factors to be discussed here are the influence of (a) the matrix, (b) the cationization agent, (c) the laser power, (d) the type of analyzer, and (e) the polydispersity of the sample. A comprehensive overview on this topic is given, for example, by Belu et al. [34].

3.4.1 Influence of the Matrix

The proper choice of the matrix is crucial for obtaining MALDI mass spectra. An overview on polymer classes and the appropriate matrices taken from Räder and Schrepp [33] is given in Table 3.4. A demonstration of the influence of the matrix on the quality of the spectra is given in Fig. 3.4. Here PEG has been investigated in different matrices. It can be recognized that the matrix giving the best signal-to-noise in this case is dithranol. Table 3.4 may serve as a first guide for a proper choice of the matrix.

Table 3.4. Overview of polymers investigated by MALDI MS and appropriate matrices. (Reprinted from [33] with permission of WILEY-VCH Verlag, Germany)

Polymer	Appropriate matrices
PMMA	HABA, DHB, IAA, dithranol
PS	HABA, dithranol, all-*trans* retinoic acid, 2-NPOE
Polystyrene sulfonic acid	SA
PEG	SA, HABA, DHB, dithranol, IAA
Polyamide-6	HABA
Polycarbonate	HABA
Polylactide	DHB, THAP
Aliphatic polyesters	HABA
PDMS	DHB, dithranol
Polybutadiene, polyisoprene	IAA, dithranol POPOP
Phenolic and epoxy resins	DHB, dihranol, THAB

3.4 Experimental Parameters

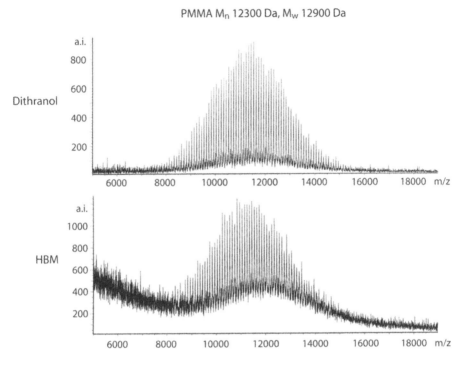

Fig. 3.4. Influence of the matrix on the MALDI mass spectrum of PMMA 12,000; solvent: THF, cationizing agent: Na-TFA; upper part: matrix dithranol; lower part: matrix HBM, all other conditions equal. The differences in S/N-ratio especially in the high mass range and the shift in baseline are obvious

3.4.2 Influence of the Cationizing Agent

The choice of the cationization agent can have the following consequences: it allows the recording of spectra (see for example the case of PS); the signal-to-noise ratio increases; a proper choice can suppress signals caused by other salts present in the sample. An overview on alkali and silver cationization for different polymers is given in Table 3.5.

A comparison between Na- and Cs-ions for the cationization of PEG is given in Fig. 3.5. The major oligomer series corresponds to the $[M+Cs]^+$ molecular ions while the oligomer series of lower intensity is due to the $[M+Na]^+$ molecular ions. This clearly indicates that Cs has a stronger affinity towards PEG as compared to Na.

The polarity of the macromolecules is of prime importance for the selection of a suitable cationization agent. Llenes and O'Malley [36] showed that the Lewis acid and base theory could be a useful guide. Most neutral polymers may be considered as bases which can share an electron with a suitable acid (or a cation).

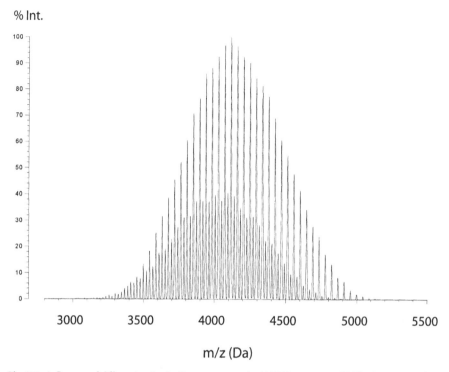

Fig. 3.5. Influence of different cationization agents on the MALDI spectrum of PEG 4000, matrix: dithranol, cationizing agents: NaCl and CsCl in equal amounts. (Reprinted from [35] with permission of DKI, Darmstadt, Germany)

Table 3.5. Metal salts and their influence on cationization, taken from [34]

Polymer	Metals that cationize polymer chains when added to the matrix	
	Ag	Alkali metals (Li, Na, K)
PS	Yes	No
Polybutadiene	Yes	NA[a]
Polyisoprene	Yes	NA
Polyethylene	No	NA
PEO	No	Yes
PMMA	Yes	Yes
PDMS	Yes	Yes

[a] Not attempted

Weak bases like PS can act as electron donor and are easily polarizable, the contrary being true for hard bases like PEG. Weak acids like silver cations carry free valence electrons and can be polarized easily. Strong acids like the alkali metal ions carry a high positive charge density and have no non-binding electrons in their valence shell. According to the Lewis acid/base concept stable acid/base pairs are formed through the combination of hard acids with hard bases and soft acids with soft bases, correspondingly. In the case of MALDI MS this means that polymers like PEG, PMMA, PA with electronegative elements like oxygen or nitrogen are best cationized by Li^+ or Na^+; unpolar polymers with π-electrons like PS or PB prefer larger polarizable cations like Ag^+ or Cu^+.

Thomson et al. [18] gave an experimental confirmation of these principles. Ionization probabilities differ for distinct chemical structures or polarities. This may lead to the fact that chains with different endgroups but similar amounts in the sample are detected with different signal intensities. This clearly hampers a quantitative analysis by MALDI MS. A typical example are MALDI mass spectra of amines where quaternary structures naturally appear as the most intense peaks followed by tertiary amines and so forth [37]. Llaine et al. [38] state in their paper on polyesters with carboxyl endgroups that both the salts and the matrices used in sample preparation strongly effect the quantitative result of an endgroup analysis.

3.4.3 Influence of the Laser Power

In Sect. 3.1 it was shown that the laser radiation plays an essential role in the ionization of the molecules and consequently should have an influence on the appearance of the mass spectra. Experimentally it has been found that a certain threshold value is necessary for the generation of meaningful mass spectra. An example is shown in Fig. 3.6. This threshold value is fairly sharp, since the ion production falls with the fifth power of laser irradiance [39]. As can be inferred from the figure above, a too high laser irradiance leads to a degradation of the mass spectra with respect to resolution and shape of the molar mass distribution. As the internal energy of the analyte molecules is raised an increasing amount of degradation can be observed [34]. So laser intensity and homogeneity of the beam profile at the sample surface have to be controlled carefully, a laser irradiance of ~20% above threshold being a good rule of thumb.

3.4.4 Type of TOF Analyzer

In addition to the parameters discussed already, the type of analyzer and extraction mode also influence the obtainable mass

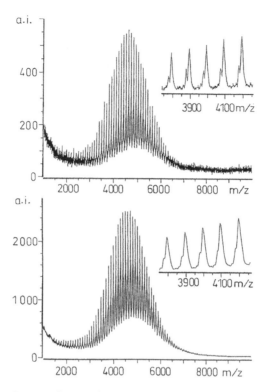

Fig. 3.6. Influence of the laser power on the MALDI mass spectrum of PMMA 4500, showing an increase in the signal-to-noise ratio and a decrease in the mass resolution, matrix: dithranol, solvent: THF, cationization agent: potassium trifluoroacetate; Laser fluence is doubled (*lower trace*) from threshold. (Reprinted from [33] with permission of WILEY-VCH, Germany)

spectra. In the beginning of the MALDI application to polymers mainly linear instruments with acceleration voltages up to 35 kV have been used. Due to the limited resolution of linear TOF analyzers in the higher mass range only the envelope of the molar mass distribution could be obtained. On the other hand the highest mass range is obtained in linear instruments. An example is given in Fig. 3.7.

These spectra in general are not totally unequivocal. A priori it is not certain if the species on the low molar mass side of the main peak result from the sample itself or represent doubly charged species from the main distribution (which is the case in the example given above). On the high molar mass side of the main distribution, clusters of the analyte molecules in form of dimers and trimers can appear. Through the proper choice of matrix, laser irradiance, analyte concentration, etc., these effects can be minimized but not prevented completely. In the linear mode the spec-

3.4 Experimental Parameters

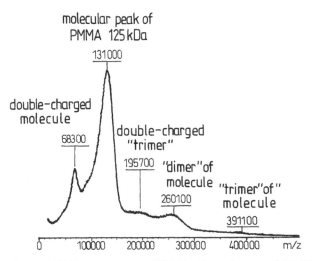

Fig. 3.7. MALDI mass spectrum of PMMA with a molar mass of 125,000 g/mol

tra with the highest molar mass values (1.5 Mio. Da for PS, see [6]) have been obtained and this mode is used for the calibration of SEC by analyzing the corresponding SEC fractions (see Chap. 6 and [40]).

With the introduction of reflectron TOF analyzers and the recovery of pulsed (or delayed) extraction techniques for MALDI spectrometers a greatly enhanced mass resolution was achieved. Delayed extraction techniques – also called time-lag-focussing – date back to work of Wiley and McLarren in 1955. They have been applied in MALDI-TOF by Spengler and Cotter [41] and to polymer analysis by Whittal et al. [42]. Figure 3.8 presents the progress in resolution exemplified for polystyrene. As can be seen, single oligomer resolution up to a mass range of 25,000 amu is possible when delayed extraction is applied.

The conceptual strength of the delayed extraction technique lies in the fact that due to the higher mass resolution different endgroups can be separated. In the lower mass range (up to 5000 amu) even single mass (isotope) resolution is possible. Also, for the characterization of the chemical heterogeneity of copolymers, maximum resolution is required to resolve all structural peculiarities. An example for a proper endgroup analysis is shown in Fig. 3.9. Here the delayed extraction technique leads to a spectrum allowing the recognition of different endgroups while direct extraction does not give useful results. Figure 3.9 gives a clear-cut example of how the effects of the correction of the initial velocity of the analyte molecules and the higher amount of extracted ions translate into a considerably better quality of the spectra or even into a useful spectrum at all.

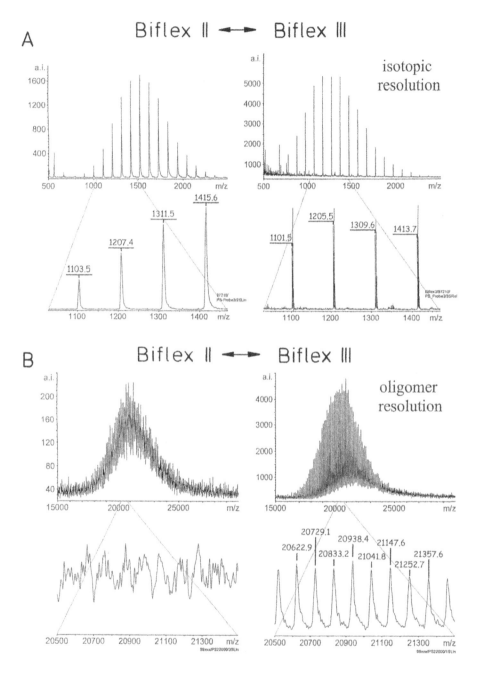

Fig. 3.8 A,B. MALDI mass spectra of PS demonstrating the effect of delayed extraction on mass resolution: **A** PS 1400; **B** PS 22000, *left spectrum* in A and B: linear TOF analyzer Biflex II, *right spectrum* in A and B: reflectron with delayed extraction Biflex III

3.4 Experimental Parameters

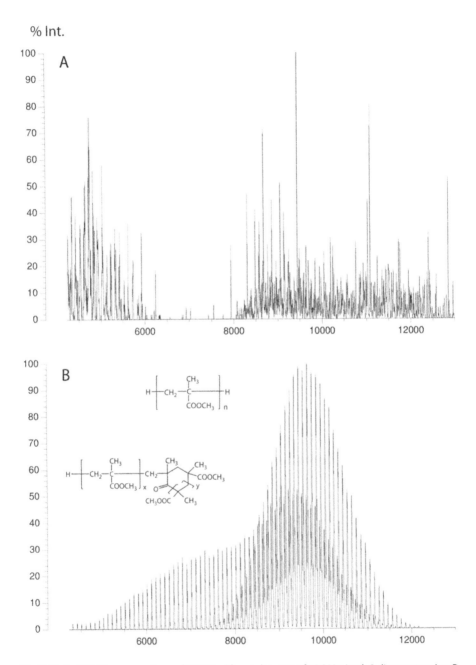

Fig. 3.9 A,B. MALDI mass spectrum of PMMA with a molar mass of 10,000 g/mol: **A** direct extraction; **B** delayed extraction of the ions. (Reprinted from [35] with permission of DKI, Darmstadt, Germany)

3.4.5 Molar Mass Distribution

In the literature there are various reports on the good coincidence between MALDI-determined M_n and M_w values for lower molar mass polymers (M_p of several thousand Da) with low polydispersity (about 1.05). Examples can be found amongst others in [34,43]. However, it soon turned out that for higher molar mass polymers and/or higher polydispersities difficulties would appear. A first indication can be seen in Fig. 3.7 which demonstrates that additional features in the spectra may occur. Figure 3.10 taken from [44]

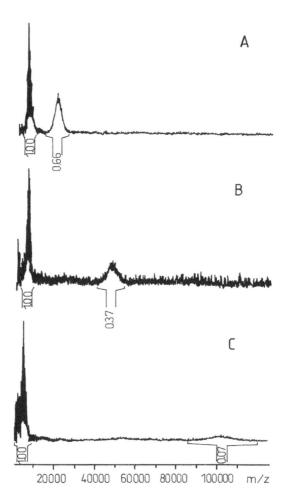

Fig. 3.10 A–C. MALDI mass spectra of equimolar mixtures of PS standards showing the influence of the molar masses on the peak areas; **A** 5500 Da + 20,800 Da; **B** 5500 Da + 46,000 Da; **C** 5500 Da + 98,000 Da; matrix 9-nitroanthracene, the small numbers give the normalized peak areas. (Reprinted from [44] with permission of Wiley, UK)

3.4 Experimental Parameters

points to the fact that for bimodal distributions the spectra are influenced by the distribution itself, the matrix, and the cation.

The influence of the laser power on the intensity ratio of a bimodal, equimolar mixture of PEGs is shown in Fig. 3.11. It reveals that the intensity ratio is a priori not 1:1 and higher molar mass oligomers need higher laser power for ionization. Systematic approaches to the influence of laser power and molar mass distribution on the MALDI spectra are reported, for example, by Martin et al. [44], Jackson et al. [45], and McEwen et al. [46].

Martin et al. mixed equimolar amounts of PS or PMMA standards in a mass range between 5000 and 100,000 Da with increasing difference in molar masses. The authors found that higher molar mass species require higher laser power for the desorption/ionization process giving less intense peaks for the high mass component and diminishing the low molar mass part of the mixture due to increased fragmentation. The dependence on laser power is different for PS and PMMA. For the mixtures they found no definite threshold value. 9-Nitroanthracene as matrix was found to have a rather high threshold for samples with a narrower oligomer distribution and thus to be less selective towards higher and lower molar mass portions than dithranol.

The lower detection efficiency for bigger ions also plays a role. The point of detector saturation was considered in detail by McEwen et al. [46]. They investigated a PMMA sample with M_p of 17,000 Da and a polydispersity of 1.8. In the SEC separation, polymers up to 40,000 Da could be detected. MALDI MS in the linear mode only delivered a more or less simple decaying distribution. The reflectron spectrum shown in Fig. 3.12 resembles the SEC curve more closely but the maximum of the masses detectable was around 20,000 Da.

By gating the signal in such a way that only a narrow bunch of ions in a selected mass range could reach the detector, the authors demonstrated that ions above 25,000 Da can be detected from the original sample. The conclusion is that the MALDI process is capable of producing ions over the entire molar mass distribution but the high mass tail cannot be detected when the lower mass share is allowed to reach the detector. So detector saturation is identified as another reason for the high-mass discrimination in polydisperse samples. It is well known that deflecting the lowest mass region of a spectrum (at least the matrix and matrix-cluster ions) can enhance the signal-to-noise ratio and mass range in a MALDI spectrum.

A final consolatory example is given in Fig. 3.13 which is typical for a broad polymer distribution (polydispersity >3). The M_w value of this sample is around 15,000 Da with the low molar mass

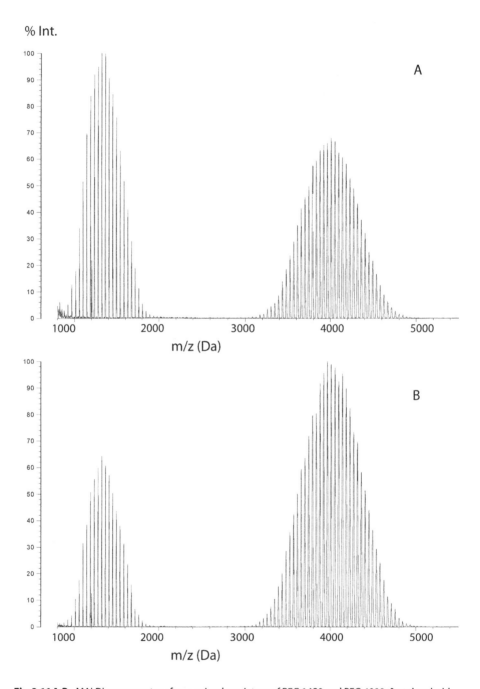

Fig. 3.11 A,B. MALDI mass spectra of an equimolar mixture of PEG 1450 and PEG 4000; **A** at threshold; **B** 10% above threshold. (Reprinted from [35] with permission of DKI, Darmstadt, Germany)

3.4 Experimental Parameters

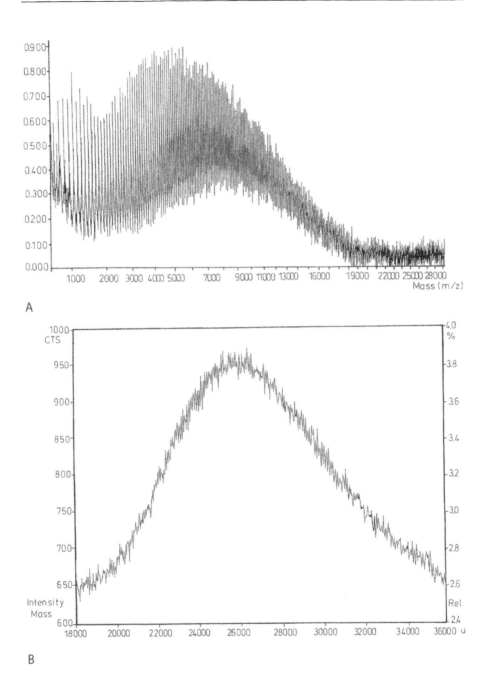

Fig. 3.12. Reflectron mass spectrum of a PMMA sample with a molar mass of 17,000 g/mol and a polydispersity of 1.8 (*top*), same sample with a parent ion selector set to transmission of ions at mass 28,000 with a 4000 Da window (Reprinted from [46] with permission of Elsevier, The Netherlands)

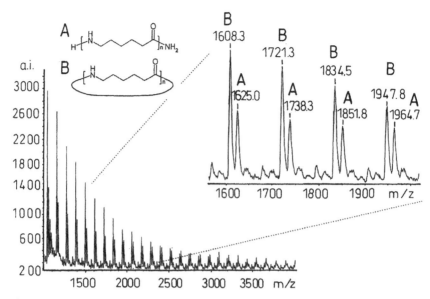

Fig. 3.13. MALDI mass spectrum of commercial polyamide-6 with a molar mass of 15,000 g/mol

tail reaching down to around 1000 Da. From Fig. 3.13 it can be inferred that the molar mass distribution is reproduced completely incorrectly – MALDI records only the low molar mass tail of the distribution – but nevertheless the repetition unit and different endgroups can be recognized.

A possible solution for this problem is to produce samples with sufficiently low polydispersity by fractionation methods like SEC, LC-CC or others. A variety of examples, mainly under the aspect of MALDI MS being a valuable absolute mass detector, is described in Chap. 7.

References

1. COTTER RJ (1987) Anal Chim Acta 195:45
2. KARAS M, BACHMANN D, BAHR U, HILLENKAMP F (1987) Int J Mass Spectrom Ion Proc 78:53
3. KARAS M, HILLENKAMP F (1988) Anal Chem 60:2299
4. TANAKA K, WAKI H, IDO Y, AKITA S, YOSHIDA Y, YOSHIDA T (1988) Rapid Commun Mass Spectrom 2:151
5. KARAS M, BAHR U (1990) Trends Anal Chem 9:321
6. SCHRIEMER DC, LI L (1996) Anal Chem 68:2721
7. KARAS M, GLÜCKMANN M, SCHÄFER J (2000) J Mass Spectrom 35:1
8. VERTES A, IRINYI G, GIJBELS R (1993) Anal Chem 65:2389
9. JOHNSON RE (1999) Int J Mass Spectrom Ion Proc 139:25

10. Juhasz P, Vestal ML, Martin SA (1997) J Am Soc Mass Spectrom 8:209
11. Glückmann M, Karas M (1999) J Mass Spectrom 34:467
12. Zenobi R, Knochenmuss R (1998) Mass Spectrom Rev 17:337
13. Israelachvili J (1991) Intermolecular and surface forces. Academic Press, London
14. Knochenmuss R, Leutwyler S (1988) Chem Phys Lett 144:317
15. Karbach V, Knochenmuss R (1998) Rapid Commun Mass Spectrom 12:968
16. Land CM, Kinsel GR (1998) J Am Soc Mass Spectrom 9:1060
17. Harrison AG (1992) Chemical ionization mass spectrometry. CRC Press, Boca Raton
18. Thomson B, Suddaby K, Rudin A, Lajoie G (1996) Eur Polym J 32:239
19. Lehmann E, Knochenmuss R, Zenobi R (1997) Rapid Commun Mass Spectrom 11:1483
20. Linnemayr K, Valla P, Allmaier E (1998) Rapid Commun Mass Spectrom 12:1344
21. Liu HMD, Schlunegger UP (1996) Rapid Commun Mass Spectrom 10:483
22. Dogruel D, Nelson RW, Williams P (1996) Rapid Commun Mass Spectrom 10:801
23. Fitzgerald MC, Parr GR, Smith M (1993) Anal Chem 65:3204
24. Perera IK, Perkins J, Kantarzoglou S (1995) Rapid Commun Mass Spectrom 9:180
25. Xiang F, Beavis RC (1993) Org Mass Spectrom 28:1424
26. Allwood DA, Perera IK, Perkins J, Dyer PE, Oldershaw GA (1996) Appl Surf Sci 103:231
27. Dwyer J, Botten D (1996) Am Lab 28:51
28. Axelsson J, Hoberg A, Waterson C, Myatt P, Shield G, Varney J, Haddleton DM, Derrick PJ (1997) Rapid Commun Mass Spectrom 11:209
29. Laiko VV, Moyer SC, Cotter RJ (2000) Anal Chem 72:5239
30. Skelton R, Dubois F, Zenobi R (2000) Anal Chem 72:1707
31. Glücksmann M, Krüger R, Pfenninger A, Thierolf M, Karas M, Hoorneffer V, Hillenkamp F, Strupat K (2001) Int J Mass Spectrom 210:121
32. Wei J, Buriak J, Siuzdak G (1999) Nature 401:243
33. Räder HJ, Schrepp W (1998) Acta Polym 49:272
34. Belu AM, DeSimone JM, Linton RW, Lange GW, Friedman RM (1996) J Am Soc Mass Spectrom 7:11
35. Ghahary R (1998) Dissertation, University of Technology, Darmstadt

36. LLENES CF, O'MALLEY (1992) Rapid Commun Mass Spectrom 6:564
37. WOLF R (2002) BASF Aktiengesellschaft, Ludwigshafen, private communication
38. LLAINE O, ÖSTERHOLM H, JÄRVINEN H, WICKSTRÖM K, VAINIOTALO P (2000) Rapid Commun Mass Spectrom 14:482
39. ENS WE, MAO Y, MEYER R, STANDING KG (1991) Rapid Commun Mass Spectrom 5:117
40. MONTAUDO G, MONTAUDO MS, PUGLISI C, SAMPERI F (1997) Int J Polym Anal Char 3:177
41. SPENGLER B, COTTER RJ (1990) Anal Chem 62:793
42. WHITTAL RM, LI L, LEE S, WINNICK MA (1996) Macromol Rapid Commun 17:59
43. SPICKERMANN J, MARTIN K, RÄDER HJ, MÜLLEN K, SCHLAAD H, MÜLLER AHE, KRÜGER R-P (1996) Eur Mass Spectrom 2:161
44. MARTIN K, SPICKERMANN J, RÄDER HJ, MÜLLEN K (1996) Rapid Commun Mass Spectrom 10:1471
45. JACKSON C, LARSEN B, McEWEN C (1996) Anal Chem 68:1303
46. McEWEN CN, JACKSON C, LARSEN B (1997) Int J Mass Spectrom 160:387

4 Identification of Polymers

4.1 Introduction

The type of polymer can be identified by determining the mass number of the repeat unit (r.u.) in the mass range where single oligomer resolution is achieved. This mass number is characteristic for a certain chemical composition and in most cases assignment is unambiguous. Very frequently, the type of polymer is already known before running a MALDI-TOF experiment. In these cases, the experiment aims at the analysis of possible by-products and various endgroups, and the determination of molecular heterogeneity. An overview on polymeric systems investigated so far by MALDI-TOF MS is given in Table 4.1.

In the following sections a number of applications for various polymers will be described in more detail. The aim of these applications is the elucidation of the chemical composition of the samples. Mostly industrial samples are selected to exemplify typical tasks for MALDI-TOF.

Table 4.1. Polymers investigated by MALDI-TOF MS

Polymer	Molar Mass	Matrix/solvent	Remarks	Ref.
PMMA	95000	HABA/THF	Variation of polydispersity	1
	18000	DHB/THF	Detection of cyclic oligomers	2
	10000	DHB/THF	Comparison to SEC	3
	15100	DHB/THF	Comparison to SEC	4
	20 and 50mer	IAA, HABA, DHB/ CHCl$_3$, acetone	No mass discrimination observed	5
	4100	DHB/methanol	Influence of pH	6
	140000	DHB/aceton	Catalytic chain transfer under emulsion conditions	7
	up to 3800	DHB/THF	Influence of cations	8

Table 4.1. (continued)

Polymer	Molar Mass	Matrix/solvent	Remarks	Ref.
	25000	IAA/acetone	Coupling with GPC	9
	260000	IAA/acetone		10
	1690/5220-mixture	Dithranol/THF	Influence of molar mass distribution	11
	~ 1000	DHB/acetone	Backbiting	2
PS	52000	HABA/THF	Variation of polydispersity	1
	12000	NA +AgTFA	Comparison to SEC	3
	1150 (syndiotactic)	NA/acetone, water	Different endgroups	13
	up to 7500	Dithranol/AgTFA	Variation of cation	8
	mixtures	Dithranol/AgTFA	Influence of cation; time-lag focusing	14
	40000	various	Comparison of different matrices	15
	125000	IAA/Ag(acac)		16
	1500000	all-trans retinoic acid	Highest polymer distribution recorded up to now	17
PS-sulfonic acid	43000	SA	Comparison of acid, salt	18
	3700–91600	SA, DHB/water	Polymer analogous sulfonation	19
PEG	23600	HABA/THF	Varying polydispersitiy	1
	35000	HABA/THF	Comparison of different matrices	20
	5000	DHB/THF	Comparison to GPC	21
	11000	Dithranol/CHCl$_3$		15
	mixture	Dithranol/HFIP	Influence of cation; time-lag focusing	14
	600	IAA/THF	After reaction with chloroacetic acid, detection of 3 distributions	22
	alkoxy terminated	DHB/water	Separation by SEC	23
	fluorescence labeled	HABA/1,4-dioxane	Time-lag focusing	24

Table 4.1. (continued)

Polymer	Molar Mass	Matrix/solvent	Remarks	Ref.
Nylon 6	3000–6000	HABA/THF	Varying polydispersity	1
	3000–6000	HABA/TFE	Endgroup determination, determination of cyclics	25
Polycarbonate	17000	HABA/THF	Varying polydispersity	1
Polylactides	1000	DHB, THAP/THF		26
Polybutylenadipate	4000	HABA/THF	Varying polydispersity	1
	39000	HABA, DHB/THF	GPC fractions	1
Polycaprolactone	10000	HABA/THF	Varying polydispersity	1
Poly THF	2000	THAP/THF		22
Polybutylmethacrylate	10000	IAA/acetone	Comparison with GPC, laser light scattering	27
Polyesteralcohol	2500	DHB/THF		22
PDMS	21000	DHB/THF	GPC-fractions	28
	6000		HPLC-separation of mixtures, detection of cyclics	29
	5000	Dithranol/AgTFA, CHCl$_3$		15
Polybutadiene	10000	POPOP/Ag(acac)		16
	5000	Dithranol	Comparison of different matrices	15
Polyisoprene	10000	POPOP/Ag(acac), THF		16
	5000	IAA/Ag, acetone	Comparison of different matrices	15
Polyesters	several 1000 (PEA; PBA; PTS)	IAA/acetone	Acidolysis	30
	(aromatic; aliphatic)	DHB	Cyclic structures, differences to SEC	21
	(PET)	Dithranol/AgTFA THF	Influence of cation, time-lag focusing	14
	9000 (adipic acid ester)	Dithranol/AgTFA THF	Transesterification with 1,4-butandiol	31

Table 4.1. (continued)

Polymer	Molar Mass	Matrix/solvent	Remarks	Ref.
Copolymers				
PS-*block*-p(α-methylstyrene)		IAA/THF	Information on both constituents	32
Tri-*block* (EO)*b*(PO)b(EO)			After LC-separation	23
Poly(buthyleneadipate-*co*-buthylenesuccinate (PBAS)		DHB, HABA/THF	Fractions from GPC	9
Bisphenol-A copolyester	64000		Correction procedure for M_w-determination	33
Poly(MMA-*co*-styrene)		DHB/acetone, water	Chain transfer reaction	7
Poly(styrene-*co*-methylacrylate)		Dithranol/AgTFA, THF	Rate coefficients	12
Poly(MMA-*co*-*n*-BMA)	2000	DHB/Na-salt, THF	Chain length distribution	34
Poly[(*o*-cresyl glycidylether)-*co*-formaldehyde]		1,4 diphenyl-butadiene	Time-lag focusing	35
Glycidylend-capped poly[(bisphenol-A)-*co*-epichlorhydrin]		HABA/dioxane	Time-lag focusing	35
PNVP/PVAc	1700	IAA/acetone	High resolution reflectron spectrum	36
Some Special Compounds				
Lignin	2600 (maximum)	DHB/acetone		37
Technical waxes	3000	2-NPOE/AgNO$_3$, xylene	Results good at intermediate masses	38
Phenolic resin novolacs	several 1000	DHB/acetone	Detailed chemical composition	39
Epoxy novolacs	several 1000	Dithranol/LiCl, THF	Detailed chemical composition	39

4.2 Analysis of Homopolymers

4.2.1 Identification of Low Molar Mass Homopolymers

Samples of molar masses below about 15,000 Da and narrow polydispersity can be measured easily using standard MALDI-TOF equipment. It has been shown by different authors that for such samples dithranol can be used as a rather universal matrix. Therefore, it makes sense to carry out a first shot MALDI-TOF experiment to identify an unknown polymer sample using dithranol as the matrix. — Aim

The samples under investigation were laboratory samples of Deutsches Kunststoff-Institut, Darmstadt, Germany. — Materials

MALDI-TOF system — Kratos Kompact MALDI III, acceleration voltage of positive 20 kV, detection in linear mode, 100–200 laser shots were summed per full spectrum — Equipment

MALDI-TOF sample preparation — a matrix solution was prepared by dissolving dithranol in THF (10 mg/ml). The sample under investigation was dissolved in THF (3–5 mg/ml) and 10 µl of the sample solution were mixed with 10 µl of the matrix solution. To the resulting mixture another 1 µl of a 0.1 mol solution of the cationizing salt was added. Then 0.5–1 µl of the final solution of sample, matrix, and salt were deposited on the sample target and air-dried.

The MALDI-TOF spectra of three different polymer samples are shown in Fig. 4.1. As can be seen from Fig. 4.1, the samples exhibit average molar masses of about 1000 to 2000 g/mol and homogeneous oligomer distributions. An inspection of the peak-to-peak mass increments reveals that they are 106 Da for sample A, 254 Da for sample B, and 192 Da for sample C. These masses correspond to the masses of the polymer repeat units, which can be assigned to phenolic novolacs (A), bisphenol-A polycarbonate (B), and polyethylene terephthalate (C), correspondingly. Accordingly, these polymers can be identified as being measured in spectra A, B, and C. — MALDI-TOF analysis

In all cases dithranol is used as the matrix. As cationizing salts are required for efficient ionization, a variety of different salts have been tested. The best results are obtained with sodium trifluoroacetate for A, lithium chloride for B, and sodium chloride for C. Accordingly, the mass peaks can be assigned to the following molecular ions:

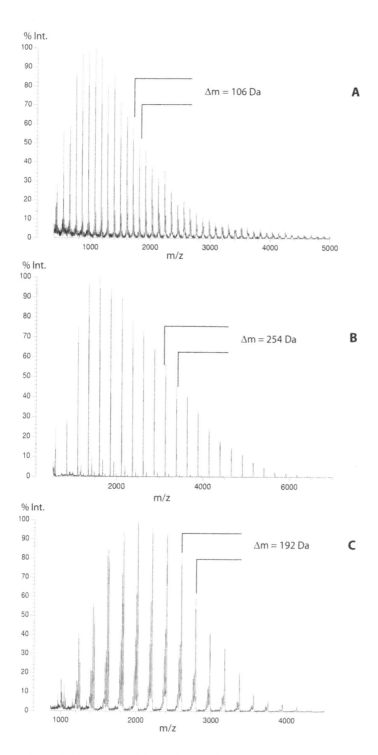

Fig. 4.1. MALDI-TOF spectra of three different polymer samples (see text), matrix: dithranol, linear mode

4.2 Analysis of Homopolymers

(A) PF novolac   [M+Na]+ = 23 (Na) + 200 (endgroups) + 106 n
(B) Polycarbonate [M+Li]+ = 7 (Li) + 326 (endgroups) + 254 n
(C) PET   [M+Na]+ = 23 (Na) + 62 (endgroups) + 192 n

4.2.2 Polysulfones

Polysulfones are used as technical polymers and exhibit high chemical and temperature stability. The end-use properties are determined by molar mass distribution and chemical composition. Due to the high polydispersity of the sample ($M_w/M_n > 2$) the recording of the true oligomer distribution is not possible. MALDI-TOF shall be used for the analysis of the different chemical species that are present in the sample. — Aim

The sample was synthesized by BASF Aktiengesellschaft, Ludwigshafen. — Materials

MALDI-TOF system — Bruker Biflex, linear mode, acceleration voltage 30 kV, 100 laser shots were summed in the linear mode per full spectrum. — Equipment

MALDI-TOF sample preparation — a matrix solution was prepared by dissolving dithranol in THF (20 mg/ml). The polysulfone was dissolved in THF (10 mg/ml). A 0.1 molar solution of K-trifluoroacetate (K-TFA) in THF was prepared. Then 10 µl of the sample solution were mixed with 10 µl of the matrix solution and 1 µl of the salt solution. Following this, 1 µl was spotted on the sample target (stainless steel) and air-dried.

The polysulfone is prepared according to the following reaction scheme: — MALDI-TOF analysis

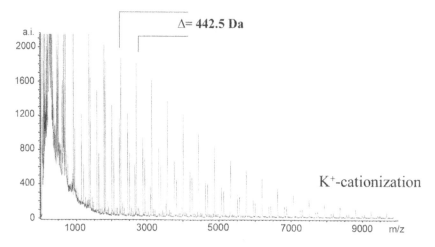

The corresponding mass spectrum is shown in Fig. 4.2. As can be inferred from Fig. 4.2, a clearly resolved oligomer distribution is obtained, displaying the expected repeat unit for polysulfone of 442.5 Da. The zoomed part of the spectrum in Fig. 4.3 shows at least four different oligomer distributions that can be assigned tentatively to the following structures:

Fig. 4.2. MALDI-TOF spectrum of polysulfone

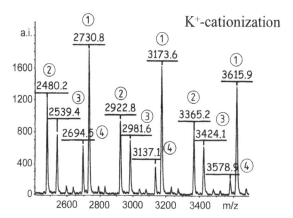

Fig. 4.3. Zoomed part of the MALDI-TOF spectrum and proposed oligomer structures fitting the peak series detected in the spectrum

4.2 Analysis of Homopolymers

(1) Cl—[—⟨O⟩—SO$_2$—⟨O⟩—O—⟨O⟩—C(CH$_3$)(CH$_3$)—⟨O⟩—O—]$_n$—H

(2) HO—⟨O⟩—C(CH$_3$)(CH$_3$)—⟨O⟩—O—[—⟨O⟩—SO$_2$—⟨O⟩—O—⟨O⟩—C(CH$_3$)(CH$_3$)—⟨O⟩—O—]$_n$—H

(3) Cl—[—⟨O⟩—SO$_2$—⟨O⟩—O—⟨O⟩—C(CH$_3$)(CH$_3$)—⟨O⟩—O—]$_n$—⟨O⟩—SO$_2$—⟨O⟩—Cl

(4) [—⟨O⟩—SO$_2$—⟨O⟩—O—⟨O⟩—C(CH$_3$)(CH$_3$)—⟨O⟩—O—]$_n$ (cyclic)

As can be seen, not only cyclic structures but also linear structures with different endgroups are identified. Nevertheless one has to keep in mind that the coincidence of the calculated mass with the one found under the resolution conditions provided by TOF spectrometers is not a strict proof though it is very likely and generally accepted. A strict proof necessitates another independent information either in form of a "correct" mass (mass resolution of 10^4 to 10^5 is required in this case) or by other spectroscopic methods like NMR or FTIR. Other mass spectroscopic possibilities like tandem mass spectrometry, collision induced dissociation, and the so-called post source decay will be described later.

4.2.3 Polyester Diols

Polyester polyols are important precursors for the production of polyurethanes. They react with diisocyanates to form linear polymer chains. If the functionality of the polyester is less than 2 it cannot be properly incorporated into the polymer chain. A functionality of 1 of the polyester causes chain termination, a functionality of 0 as in cyclic oligomers causes the presence of species that are non-reactive. Therefore, the diagnosis of cyclics as by-products is of key importance for the evaluation of the quality of the polyester diol. — *Aim*

The polymer was synthesized by BASF Aktiengesellschaft, Ludwigshafen. — *Materials*

Equipment

MALDI-TOF system — Bruker Biflex II, acceleration voltage 20 kV, detection in the reflectron mode, 100 laser shots accumulated.

MALDI-TOF sample preparation — HBM(4-hydroxy-benzylidene malonitrile) dissolved in THF (20 mg/ml) was used as matrix. The sample was dissolved in THF (10 mg/ml). As cationizing agent a 0.1 molar solution of K-TFA in THF was used. The solutions were mixed 10:10:1 (analyte:matrix:salt) and 1 µl spotted onto the steel target and air-dried.

MALDI-TOF analysis

The spectrum of the polyester diol shown in Fig. 4.4 reveals oligomer distributions of at least two different species in the lower mass range. Both exhibit a peak-to-peak mass difference of 172.2 Da that is equivalent to the repeat unit of the polyester:

(1) $HOCH_2CH_2O-[CO(CH_2)_4CO-OCH_2CH_2O]_n-H$

$n=8$
$[M+K]^+ = 101 + 172.2n$
$[M+K]^+ = 1478.7$ Da (calc.)
$[M+K]^+ = 1478.4$ Da (exp.)

(2) cyclic $[CO(CH_2)_4CO-OCH_2CH_2O]_n$

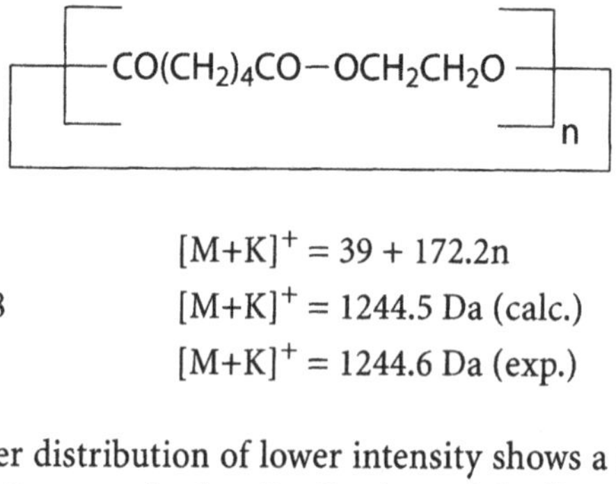

$n=8$
$[M+K]^+ = 39 + 172.2n$
$[M+K]^+ = 1244.5$ Da (calc.)
$[M+K]^+ = 1244.6$ Da (exp.)

The oligomer distribution of lower intensity shows a mass difference of −62 Da towards the distribution of the linear oligomers thus corresponding to cyclic oligomers through the loss of one ethylene glycol molecule.

Figure 4.4 shows a remarkable trend for the cyclic oligomers: in general the cyclic oligomers are only found in the lower mass region of the oligomer distribution which is reasonable as the probability for the formation of cyclic species should diminish with increasing molar mass.

4.2 Analysis of Homopolymers

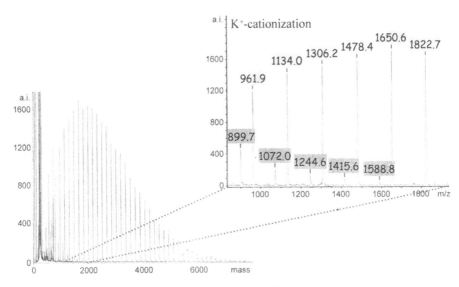

Fig. 4.4. MALDI mass spectrum of the polyester diol prepared from ethylene glycol and adipic acid

4.2.4 Poly(oxymethylene)

Poly(oxymethylene) (POM) or polyformaldehyde, mainly comprised of oxymethylene units -CH_2O-, is used as an engineering plastic. It is most frequently synthesized by anionic polymerization of formaldehyde or cationic polymerization of 1,3,5-trioxane. In both cases POM molecules are formed that mainly have hydroxy endgroups. Because POM molecules with these endgroups tend to depolymerize upon heating, the stability of POM is improved by end-capping via esterification or etherification. Therefore, the structure of the endgroups is of interest for the stability behavior and shall be determined by MALDI-TOF. — Aim

The polymer was synthesized by BASF Aktiengesellschaft, Ludwigshafen. — Materials

MALDI-TOF system — Bruker Biflex II, acceleration voltage 20 KV, detection in the reflectron mode, 100 laser shots accumulated. — Equipment

MALDI-TOF sample preparation — Dithranol in 1,1,1,3,3,3-hexafluoro-2-propanol (HFIP) (20 mg/ml) was used as matrix, the polymer was dissolved in HFIP (10 mg/ml), the cationization agent K-TFA 0.1 molar in THF. Then 10 µl of matrix solution were mixed with 10 µl of polymer solution and 1 µl of salt solution. About 1 µl was spotted on the steel target and air-dried.

The MALDI-TOF mass spectrum of the poly(oxymethylene) is shown in Fig. 4.5. The enlarged part of the spectrum reveals at least three different oligomer distributions for which tentative as- — MALDI-TOF Analysis

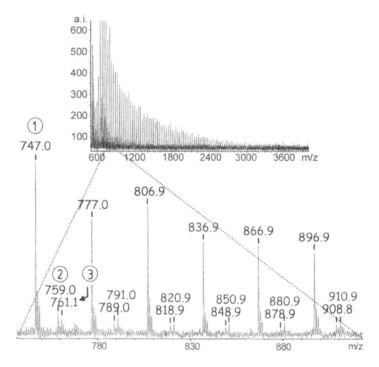

Fig. 4.5. MALDI-TOF mass spectrum of poly(oxymethylene)

signments are given. The mass of the POM repeat unit is 30 Da. No cyclic species could be detected in the present case; see also [40] for the analysis of model compounds:

(1) $H-[-O-CH_2-]_n-OH$ $[M+K]^+ = 57 + 30n$

(2) $CH_3CO-[-O-CH_2-]_n-OH$ $[M+K]^+ = 99 + 30n$

(3) $CH_3-[-O-CH_2-]_n-OH$ $[M+K]^+ = 71 + 30n$

4.2.5 Emulsifiers

Aim

Polyglycerolesters are used as emulsifiers. Starting materials are ricinoic acid, ethylene oxide (EO) and glycerol. Since different reaction pathways are possible, the resulting reaction products can be quite complex in composition. Reaction products of glycerol and ethylene oxide, ethylene oxide and ricinoic acid, and glycerol, ethylene oxide and ricinoic acid can be formed. For quality control the exact composition of the reaction products is an important parameter. Therefore, MALDI-TOF shall be used to obtain information on the composition (chemical heterogeneity) of an

4.2 Analysis of Homopolymers

emulsifier formulation. An idealized general structure of the polyglycerolester can be presented as follows:

$$\begin{array}{l} CH_2-(EO)_n-R \\ CH-(EO)_n-R \\ CH_2-(EO)_n-R \end{array}$$

R:

$$-O-\overset{O}{\underset{}{C}}-CH_2-CH_2-CH_2-CH=CH-CH_2-\underset{OH}{CH}-CH_2-CH_2-CH_2-CH_2-CH_3$$

The polymer was synthesized by BASF Aktiengesellschaft, Ludwigshafen. — Materials

MALDI-TOF system — Bruker Biflex II, acceleration voltage 20 kV, 100 laser shots were accumulated. — Equipment

MALDI-TOF sample preparation — The polymer sample was dissolved in THF (10 mg/ml), as matrix a solution of dithranol in

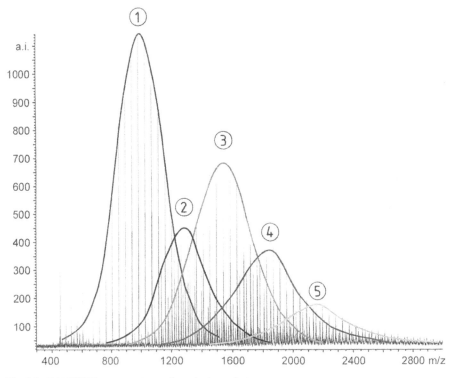

Fig. 4.6. MALDI-TOF mass spectrum of a polyglycerolester

MALDI-TOF Analysis

THF (20 mg/ml) was used, as cationizing agent K-TFA 0.1 molar in THF. 1 μl of a mixture 10:10:1 (polymer:matrix:salt) was spotted onto the steel target and air-dried.

The MALDI-TOF mass spectrum given in Fig. 4.6 reveals a complex product composition. At least five different oligomer distributions can be recognized. Experiments with PEG itself and other PEG esters suggest that in this case the oligomer distribution of the various products is displayed correctly and yields quantitative information. A plausible assignment of the five distributions is given in Fig. 4.7.

(1) HO—[CH$_2$—CH$_2$—O]$_n$—H

(2) R—[CH$_2$—CH$_2$—O]$_n$—H

(3) HO—[CH$_2$—CH$_2$—O]$_n$—CH(CH$_2$—[CH$_2$—CH$_2$—O]$_n$—OH)(CH$_2$—[CH$_2$—CH$_2$—O]$_n$—OH)

(4) R—[CH$_2$—CH$_2$—O]$_n$—CH(CH$_2$—[CH$_2$—CH$_2$—O]$_n$—OH)(CH$_2$—[CH$_2$—CH$_2$—O]$_n$—OH)

(5) R—[CH$_2$—CH$_2$—O]$_n$—CH(CH$_2$—[CH$_2$—CH$_2$—O]$_n$—OR)(CH$_2$—[CH$_2$—CH$_2$—O]$_n$—OH)

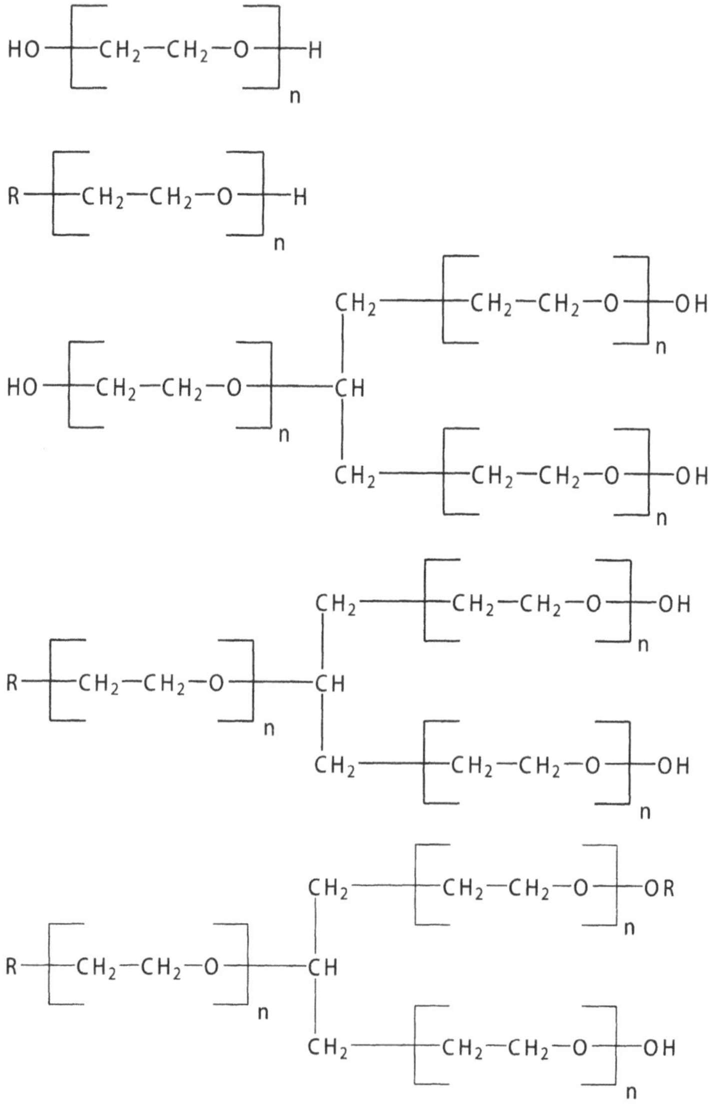

Fig. 4.7. Assignment of the oligomer distributions found in Fig. 4.6

4.2 Analysis of Homopolymers

This application exemplifies that MALDI-TOF is a very suitable method for the characterization of emulsifiers and surface active materials in general.

4.2.6 Polysiloxane

Polysiloxanes are used as additives in automotive coatings to adjust flow properties and surface hydrophobicity of the lacquers. Due to their surface migration properties, which can cause surface defects, a detailed knowledge of molar mass and chemical structure is of interest. **Aim**

Poly(dimethylsiloxane) (PDMS) (37,832-1) was purchased from Aldrich, Steinheim, Germany. **Materials**

MALDI-TOF system — Bruker Biflex II, acceleration voltage 20 kV, 100 laser shots accumulated in the reflectron mode. **Equipment**

MALDI-TOF sample preparation — The sample (10 mg/ml), the matrix dithranol (20 mg/ml) and K-TFA (0.1 molar) were dissolved in toluene. 1 µl of the mixture 10:10:1 (polymer:matrix:salt) was spotted on the steel target and air-dried.

The corresponding spectrum is given in Fig. 4.8. In the present case, a single oligomer distribution is found fitting to the structure given below with the typical repeat unit of 74 Da. The fact that **MALDI-TOF analysis**

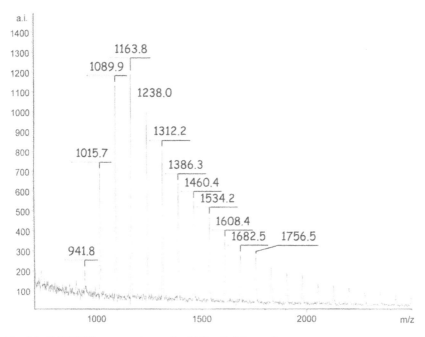

Fig. 4.8. MALDI-TOF mass spectrum of poly(dimethylsiloxane).

no other oligomer distributions are obtained is an indication for the high purity of the sample. However, this has to be confirmed by other indipendent methods:

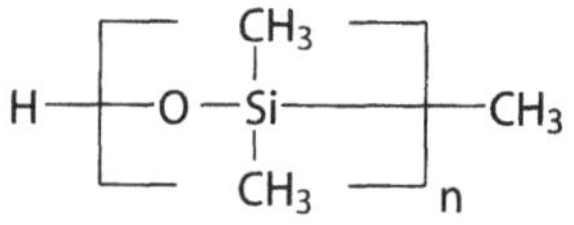

$$H{-}{\left[{-}O{-}\underset{\underset{CH_3}{|}}{\overset{\overset{CH_3}{|}}{Si}}{-}\right]_n}{-}CH_3$$

$[M+K]^+ = 55 + 74\,n$

for n = 15: $[M+K]^+ = 1165$ Da (calc.)
$[M+K]^+ = 1163.8$ Da (exp.)

4.3 Copolymers

Copolymers in general exhibit more complex spectra. Due to the various possible combinations of the monomer units, very frequently different oligomer compositions result in nearly the same mass. In such cases a very high mass resolution of the MALDI-TOF spectrometer is required.

4.3.1 Caprolactam-Pyrrolidone Copolymer

Aim

Various combinations of the monomers N-vinylcaprolactam and N-vinylpyrrolidone result in very similar oligomer masses. Due to the overlapping of the mass peaks for different oligomer compositions it is difficult to elucidate the copolymer composition. In the present application, an assignment of the mass peaks to different copolymer compositions shall be attempted. The copolymer under investigation has the following general structure:

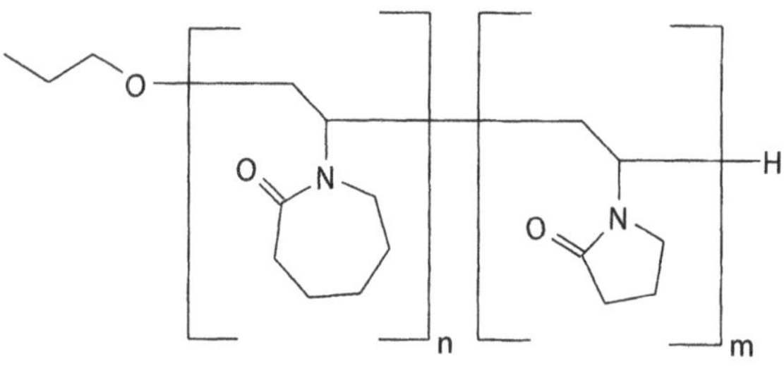

Materials

The polymer was synthesized by BASF Aktiengesellschaft, Ludwigshafen.

Equipment

MALDI-TOF system — Bruker Biflex II used in the reflectron mode, acceleration voltage 20 kV; 100 laser shots were accumulated for the spectrum displayed in Fig. 4.9.

4.3 Copolymers

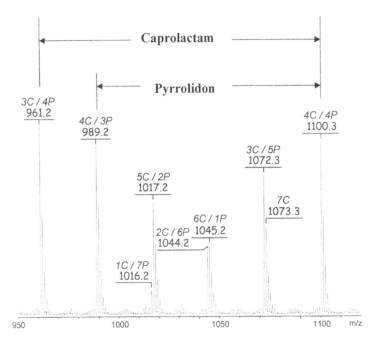

Fig. 4.9. MALDI mass spectrum of a copolymer of N-vinylpyrrolidone and N-vinylcaprolactam; C indicates a caprolactam repeat unit, P indicates a pyrrolidone repeat unit

MALDI-TOF sample preparation — Polymer (10 mg/ml), matrix dithranol (20 mg/ml) and K-TFA (0.1 molar) were dissolved in THF. 1 µl of the mixture 10:10:1 (polymer:analyte:salt) was spotted on the steel target and air-dried.

MALDI-TOF Analysis

The section of the MALDI mass spectrum given in Fig. 4.9 demonstrates that it is possible to detect oligomers which differ by only one mass unit, for example 1C+7P vs 5C+2P and 2C+6P vs 6C+1P. In this case the natural isotope distribution can also be used as an indication, if there are one or more constituents underlying an oligomer signal:

$$[M+K]^+ = 99 + 111.1\,m + 139.1\,n$$

m: number of repeat units of pyrrolidone (P)
n: number of repeat units of caprolactam (C)

for 1C+7P: $[M+K]^+$ = 1015.8 Da (calc.) / 1016.2 Da (exp.)
for 5C+2P: $[M+K]^+$ = 1016.7 Da (calc.) / 1017.2 Da (exp.)
for 2C+6P: $[M+K]^+$ = 1043.8 Da (calc.) / 1044.2 Da (exp.)
for 6C+1P: $[M+K]^+$ = 1044.7 Da (calc.) / 1045.2 Da (exp.)

Further examples on the analysis of copolymers can be found in Table 4.1 and in Chap. 6.

4.3.2 Post Source Decay

As already pointed out in Chap. 3 and as demonstrated by the examples in this chapter for homopolymers (and simple copolymers) it is possible to deduce the repeat unit of the polymer from the oligomer spectra. Also it is often possible to infer plausible endgroups but in a strict sense this conclusion is not an ultimate proof since from one signal two endgroups have to be determined. The situation quickly gets even more complicated if several endgroups can occur or the polymer itself is more complex (copolymers or terpolymers for example). Even in the case of simple copolymers it can happen that the difffferent monomers can be rather close to each other in mass or that the multiple of one monomeric unit is close to the repeat unit of the other monomer. A possible remedy is provided by Tandem Mass Spectrometry.

A technique developed for reflectron TOF analyzers delivering further fragmentation products is the so-called Post Source Decay (PSD) [41]. It makes use of the fact that a certain fraction of the desorbed analyte undergoes fragmentation/neutralization reactions during its flight time in the mass spectrometer. The activation energy originates from multiple collisions with matrix molecules during the desorption/ionization process (plume expansion, see Sect. 3.1) and with gas molecules in the drift region between accelerating grid and reflector. The activated parent (or mother) ion M^+ undergoes fragmentation according to

$$M^+ \rightarrow m_f^+ + m_n \qquad (4.1)$$

with the fragment ion m_f^+ and the neutral species m_n. Since only ions after acceleration can be detected, this occurs in the field free drift region with the consequence that all involved species continue their flight at the same speed. Due to the change in mass, the kinetic energy of the species varies significantly according to $E_{kin} = \frac{1}{2}mv^2$. As the reflectron is an energy filtering device (only ions entering the reflecting field to a certain extent are reflected properly onto the detector) the fragment ions can be detected by analyzing their kinetic energy in a reflectron instrument. The energy of the fragment ion contains only a fraction of the full acceleration energy and is proportional to the mass ratio of fragment and parent ion:

$$E_f = E_{acc}(m_f/M) \qquad (4.2)$$

4.3 Copolymers

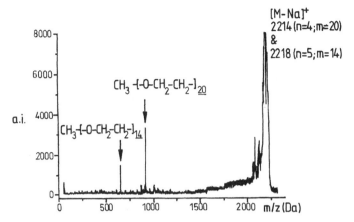

Fig. 4.10. PSD MALDI-TOF spectrum of PEO-*b*-PPE from mother ions with nearly the same molar mass (2214 Da for n=4 and m=20 and 2218 Da for n=5 and m=14). (Reprinted from [42] with permission of American Chemical Society, U.S.A.)

Only ions with sufficiently high energy can enter into the second stage of the reflectron. The ion needs to have more than approximately 50% of the parent ion energy to reach the detector. By a stepwise reduction of the reflector voltage it is possible to obtain full mass range fragment ion spectra.

An electronic gate installed in the drift tube allows one to select the mother ion(s) to be investigated and then to acquire exclusively its fragment ion spectrum. The segments recorded for each reflectron voltage are pasted together to give a complete spectrum. Pasting and calibration of the segments in general are performed by supplier software.

As an example, the analysis of a poly(ethyleneoxide)-*b*-poly(*p*-phenylene ethynylene) diblock copolymer [42] is given in Fig. 4.10. In this case two mother ions of nearly the same molar mass have been chosen and selected together. The PEO block length m from each composition can be directly evaluated from the clearly separated fragment peaks in the PSD spectrum. The expected mass of the copolymers corresponds to

$$M = (15 + 44m + 268n + 101) \text{ Da}. \qquad (4.3)$$

As the mass of the *p*-phenylene ethynylene monomer with 268 Da is almost equal to six times the ethylene oxide monomer, copolymers with different composition can give nearly the same mass (see Fig. 4.10). In this case the PSD-spectra exhibit a main cleavage between the two blocks allowing a direct identification of the PEO block length and by using Eq. (4.3) the calculation of the length of the second block.

Different copolymers such as alternating or random copolymers should generate different – and probably more complex – fragmentation patterns. PSD is also helpful in the analysis of unknown endgroups and, obviously, in cases where one monomer exhibits the multiple mass of the other one. Examples referring to other classes of polymers can be found in the literature, e.g., polycarbonate in [43] and polystyrene in [44].

References

1. Montaudo G, Montaudo MS, Puglisi C, Samperi F (1995) Rapid Commun Mass Spectrom 9:453
2. Pasch H, Gores F (1995) Polymer 36:1999
3. Lloyd PM, Suddaby KG, Varney JE, Scrivener E, Derrick PJ, Haddleton DM (1995) Eur Mass Spectrom 1:293
4. Jackson C, Larsen B, McEwen C (1996) Anal Chem 68:1303
5. Larsen BS, Simonsick WJ Jr, McEwen CN (1996) J Am Soc Mass Spectrom 7:287
6. Dogruel D, Nelson RW, Williams P (1996) Rapid Commun Mass Spectrom 10:801
7. Suddaby KG, Haddleton DM, Hastings JJ, Richards SN, O'Donnell JP (1996) Macromolecules 29:8083
8. Thomson B, Suddaby K, Rudin A, Lajoie G (1996) Eur Polym J 32:329
9. Montaudo G (1996) Polym Prep 37:290
10. Danis P, Karr DE (1993) Org Mass Spectrom 28:923
11. Spickermann J, Martin K, Räder HJ, Müllen K, Schlaad H, Müller AHE, Krüger R-P (1996) Eur Mass Spectrom 2:161
12. Schweer J, Sarnecki J, Mayer-Posner F, Müllen K, Räder HJ, Spickermann J (1996) Macromolecules 29:4536
13. Duncalf WJ, Wade HJ, Waterson C, Derrick PJ, Haddleton DM, McCamley A (1996) Macromolecules 29:6399
14. Jackson AT, Yates HT, MacDonald WA, Scrivens JH, Critchley G, Brown J, Deery MJ, Jennings KR, Brookes C (1997) J Am Soc Mass Spectrom 8:132
15. Belu AM, DeSimone JM, Linton RW, Lange GW, Friedman RM (1996) J Am Soc Mass Spectrom 7:11
16. Danis PO, Karr DE, Xiong Y, Owens KG (1996) Rapid Commun Mass Spectrom 10:862

17. SCHRIEMER DC, LI L (1996) Anal Chem 68:2721
18. DANIS PO, KARR DE (1995) Macromolecules 28:8548
19. RÄDER HJ, SPICKERMANN J, MÜLLEN K (1995) Macromol Chem Phys 196:3967
20. LIU HMD, SCHLUNEGGER UP (1996) Rapid Commun Mass Spectrom 10:483
21. BLAISS JC, TESSIER M, BOLBACH G, REMAUD B, ROZESS L, GUITARD J (1995) Int J Mass Spectrom Ion Processes 144:131
22. KRÜGER R-P (1995) GIT Fachz Lab 3:189
23. PASCH H, RODE K (1995) J Chromatogr A 699:21
24. LEE S, WINNICK MA, WHITTAL RM, LI L (1996) Macromolecules 29:3060
25. MONTAUDO G, MONTAUDO MS, PUGLISI S, SAMPERI F (1996) J Polym Sci A Polym Chem 34:439
26. KRÜGER R-P, MUCH H, SCHULZ G (1996) GIT Fachz Lab 4:398
27. DANIS PO, KARR DE, SIMONSICK WJ JR., WU DT (1995) Macromolecules 28:1229
28. MONTAUDO G, MONTAUDO MS, PUGLISI C, SAMPERI F (1995) Rapid Commun Mass Spectrom 9:1158
29. JUST U, KRÜGER R-P (1996) In Auner N, Weis J (eds) Organosilicon chemistry II. VCH, Weinheim, Germany
30. WILLIAMS JB, GUSEV AI, HERCULES DM (1997) Macromolecules 30:3781
31. CHAUDHARY AK, CRITCHLEY G, DIAF A, BECKMANN EJ, RUSSELL AJ (1996) Macromolecules 29:2213
32. WILCZEK-VERA G, DANIS PO, EISENBERG A (1996) Macromolecules 29:4036
33. MONTAUDO G, SCAMPORRINO E, VITALINI D, MINEO P (1996) Rapid Commun Mass Spectrom 10:1551
34. SUDDABY KG, HUNT KH, HADDLETON DM (1996) Macromolecules 29:8642
35. SCHRIEMER DC, WHITTAL RM, LI L (1997) Macromolecules 30:1955
36. DANIS PO, SANCY DA, HUBY FJ (1996) Polym Prep 37:311
37. BICKE C, METZGER JO (1993) GIT Fachz Lab 2:77
38. KÜHN G, WEIDNER ST, JUST U, HOLMER G (1996) J Chromatogr A 732:111
39. PASCH H, RODE K, GHAHARY R, BRAUN D (1996) Angew Makromol Chem 241:95
40. SATO H, OHTANI H, TSUGE S, HAYASHI N, KATOH K, MASUDA E, OHNISHI K (2001) Rapid Commun Mass Spectrom 15:82
41. SPENGLER B, KIRSCH W, KAUFMANN R (1992) Phys Chem 96:9678
42. PRZYBILLA L, FRANCKE V, RÄDER HJ, MÜLLEN K (2001) Macromolecules 34:4401

43. Przybilla L, Räder HJ, Müllen K (1999) Eur Mass Spectrom 5:133
44. Jackson AT, Bunn A, Hutchings LR, Kiff FT, Richards RW, Williams J, Green MR, Bateman RH (2000) Polymer 41:7437

5 Molar Mass Determination

5.1 Introduction

For polymeric materials, the molar mass or molecular size plays a critical role in determining the mechanical, bulk, and solution properties. These properties govern polymer processing, end-use performance, and stability. Therefore, determination of molar mass and molar mass distribution is of central interest in polymer analysis; see Chap. 1.

The characteristic feature of all molecules is their molar mass. Low molar mass compounds and most biopolymers are monodisperse with respect to molar mass, i.e., they have a strictly defined molar mass, which may be calculated knowing the chemical composition of the molecule. The degree of polymerization of a macromolecule P is defined as the number of repeat units within this molecule. P is related to molar mass M by $P = M/M_0$, where M_0 represents the molar mass of the repeat unit.

All synthetic polymers are polydisperse or heterogeneous in molar mass. The *molar mass distribution* originates from randomness of the polymerization process. In the daily routine synthetic polymers are often characterized by average molar masses, considering the numbers of macromolecules of a certain molar mass M_i in the sample. Most frequently used are the *number-average* molar mass M_n, expressing the amount of species by the number of moles n_i, and the *weight-average* molar mass M_w, considering the mass m_i of the species.

The difference between number and weight average molar masses gives a first estimate of the width of the molar mass distribution (MMD). The broader the distribution, the larger is the difference between M_n and M_w. The ratio of M_w/M_n is a measure of the breadth of the molar mass and is termed the polydispersity Q.

Well established methods for molar mass determination are vapor pressure and membrane osmometry, viscometry, light scattering, and ultracentrifugation; see Chap. 1. The most important method is size exclusion chromatography (SEC), also referred to as gel permeation chromatography (GPC). SEC is not an absolute method and must, therefore, be calibrated. It shows inaccuracies when samples are measured for which no calibration standards exist. This is the case for copolymers and samples with unusual architectures, like dendrimers or rigid rods. Accordingly, there is a strong need for new analytical techniques that can determine absolute molar masses.

Mass spectrometry is ideally suited for molar mass analysis due to some unique features like absolute measurement irrespective of chemical structure, very low sample consumption, and short analysis time. Out of the huge variety of different mass spectrometric techniques, MALDI-TOF mass spectrometry seems to be best suited for the analysis of polymers. This is due to the fact that MALDI-TOF allows for measurement of very high molar masses with virtually no fragmentation. In addition, at masses below 50,000 g/mol single polymer chains are resolved, enabling the determination of repeat units and endgroups. The fact that mostly singly charged ions are formed in the MALDI process makes it easy to interpret the resulting mass spectra.

Originally developed for the analysis of biomolecules in 1988, it was shown in 1992 that MALDI-TOF can be used for the analysis of synthetic polymers with molar masses above 100,000 Da [1-3]. These measurements were conducted on samples with a low polydispersity and the comparison with data from SEC showed a rather good agreement. Accordingly, from these very preliminary results expectations developed that MALDI-TOF could be used for the elucidation of molar mass distributions over a wide mass range. The expectations became even higher when Schriemer and Li reported on the detection of narrow disperse polystyrene with a molar mass as high as 1.5 million Da [4].

At the same time, however, MALDI-TOF measurements on samples with a polydispersity above 1.5 showed an insufficient agreement with SEC data. It turned out that the measurements of broadly distributed samples have to be considered with some caution. By comparing samples of varying polydispersity with MALDI-TOF and SEC it has been shown by Montaudo et al. [5] that molar mass distributions agreed only for very low polydispersities. Especially in cases where the polydispersity was larger than 2, only spectra decreasing in intensity and corresponding to the low molar mass fraction of the distribution were obtained. Only a polydispersity of <1.1 resulted in good agreement between SEC and MALDI-TOF [6].

The reasons for the discrepancy between MALDI-TOF and SEC has been investigated by a number of authors. It has been pointed out that MALDI is an ionization method which itself has considerable variation, especially in sample preparation and instrumental effects. Significant variation of polymer molar mass distributions

5.1 Introduction

is possible with TOF analyzers because of differences in ion acceleration, ion focusing, ion detection, and so forth. Thus, instrumentation or sample preparation may be the cause of different conclusions concerning molar mass analyses rather than the MALDI ionization process itself.

Since the beginnings of MALDI-TOF application for polymers it turned out that ionization and detection mechanisms have an effect on the detectable mass range and the shape of the distribution curve. These include the selection of a proper matrix, addition of salt and the sample preparation conditions (dried-drop, spin coating, spraying). The influence of different cations added to the matrix solution on the spectra of PMMA has been reported by Lloyd et al. [7]. A typical result is shown in Fig. 5.1. Using Cs^+ instead of Na^+ for cationization not only changed the intensity but also the position of the oligomer distribution. The severe effect of the matrix on the obtainable MALDI-TOF result is demonstrated in Fig. 5.2 for polystyrene. While with dithranol as the matrix a well resolved oligomer distribution is obtained, the use of indole acrylic acid (IAA) or nitrophenyl octyl ether (NPOE) results in poorly resolved spectra of low intensity [8]. A more detailed discussion of the effects of different experimental parameters on the quality of the spectra is given in Chap. 3.

The common method for determining the accuracy of a MALDI-TOF result is by comparison with values provided by SEC. These values are frequently given as M_p values. M_p is the mass

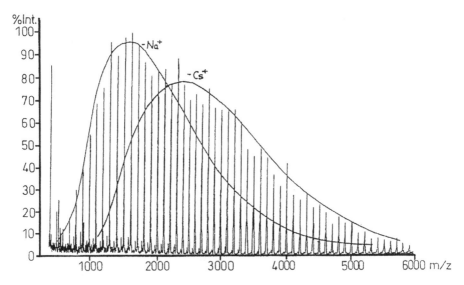

Fig. 5.1. Influence of alkali salts on the MALDI-TOF spectrum of PMMA. (Reprinted from [7] with permission of American Chemical Society, U.S.A.)

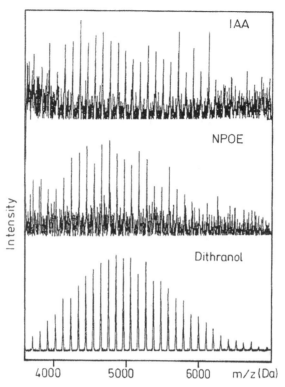

Fig. 5.2. Influence of the matrix on the MALDI-TOF spectrum of PS, silver trifluoroacetate was used as cationization agent. (Reprinted from [8] with permission of American Chemical Society, U.S.A.)

value for the most abundant point on the SEC chromatogram. Because the most abundant peak is easy to determine by mass spectrometry, it is common to compare mass spectral generated M_p values with M_p values generated by SEC [9,10]. Jackson et al. [11] demonstrated that the most probable peak value (M_p) and the shape of the distribution curve as determined by MALDI-TOF and SEC are a function of how the data are displayed.

In general, SEC results are presented as weight fractions vs the logarithm of molar mass. Plotting the mass spectral data as number fraction on a linear (or square root) mass scale has a considerable influence on the shape of the spectrum and the resulting M_p value. Figure 5.3 shows the MALDI-TOF distribution measured for a commercial poly(tetramethylene ethylene glycol).

The measured M_p by MALDI-TOF mass spectrometry is roughly 1400 while SEC gives an M_p of 4000; see Fig. 5.4A. However, plotting the SEC data as number fraction vs a square root mass scale (Fig. 5.4B) gives an M_p value of 1450, which is in close agreement with the MALDI-TOF result. Note the close similarity

5.1 Introduction

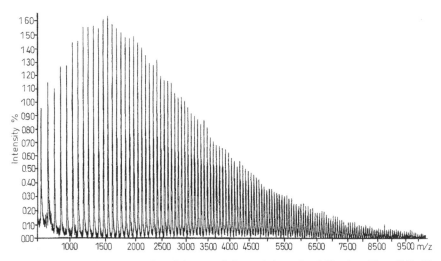

Fig. 5.3. MALDI-TOF spectrum of a poly(tetramethylene ethylene glycol) (Reprinted from [11] with permission of American Chemical Society, U.S.A.)

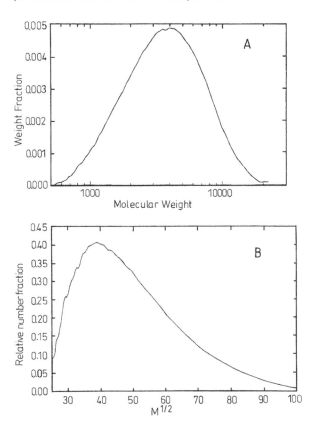

Fig. 5.4 A, B. SEC chromatogram of poly(tetramethylene ethylene glycol): **A** plotted in the usual manner; **B** plotted as number fraction vs square root of molar mass. (Reprinted from [11] with permission of American Chemical Society, U.S.A.)

of the shape of the molar mass distribution when the MS data (Fig. 5.3) and the SEC data (Fig. 5.4B) are plotted in the same manner.

5.2 Analysis of Polymers with Narrow Molar Mass Distribution

As has already been pointed out, direct and accurate molar mass determination of synthetic polymers is only possible in cases where the polydispersity of the samples is lower than roughly 1.2. The following section describes a number of applications where narrowly disperse polymer samples that are otherwise used as calibration standards for SEC are analyzed by MALDI-TOF mass spectrometry. The molar mass averages and the polydispersities calculated from the MALDI-TOF spectra are compared with results obtained by SEC.

5.2.1 Oligomer Samples

Aim

Samples of molar masses below about 15,000 Da and narrow polydispersity can be measured easily using standard MALDI-TOF equipment. It has been shown by different authors that for such samples dithranol can be used as a rather universal matrix. However, depending on the chemical composition of the sample a proper cationizing salt must be selected. As has been shown in the previous section, the cation can affect the shape of the MALDI spectrum and the position of the most abundant peak. It is, therefore, important to investigate the effect of the cation on the ionization behavior of a specific sample.

Materials

The samples under investigation were commercial narrowly disperse calibration standards for SEC. They were purchased from Polymer Standards Service GmbH (Mainz, Germany).

Equipment

MALDI-TOF system — Kratos Kompact MALDI III, acceleration voltage of positive 20 kV, detection in linear mode, 100–200 laser shots were summed per full spectrum

MALDI-TOF sample preparation — a matrix solution was prepared by dissolving dithranol in THF (10 mg/ml). The sample under investigation was dissolved in THF (3–5 mg/ml) and 10 µl of the sample solution were mixed with 10 µl of the matrix solution. To the resulting mixture another 1 µl of a 0.1 mol solution of the cationizing salt was added. Then 0.5–1 µl of the final solution of sample, matrix, and salt were deposited on the sample target and air-dried.

5.2 Analysis of Polymers with Narrow Molar Mass Distribution

The MALDI-TOF spectra of polystyrene, polymethyl methacrylate, and polyethylene glycol with average molar masses of about 4000 g/mol are presented in Fig. 5.5. In all cases dithranol is used as the matrix. As cationizing salts, silver trifluoroacetate (PS), lithium chloride (PMMA) and sodium chloride (PEO) are used, respectively. As can be seen, in all cases well resolved oligomer distributions are obtained, and the mass peaks can be assigned to the following molecular ions:

MALDI-TOF Analysis

$C_4H_9-[CH_2-CH(C_6H_5)]_n-H$ $H-[CH_2-C(CH_3)(COOCH_3)]_n-H$ $HO-[CH_2-CH_2-O]_n-H$

(A) PS $[M+Ag]^+ = 108$ (Ag) + 58 (endgroups) + 104.15 n
(B) PMMA $[M+Li]^+ = 7$ (Li) + 2 (endgroups) + 100.12 n
(C) PEO $[M+Na]^+ = 23$ (Na) + 18 (endgroups) + 44.05 n

Taking the relative intensities of the mass peaks as equivalents for the number of molecules n_i, the molar mass averages M_n and M_w, and the polydispersity M_w/M_n can be calculated from the MALDI-TOF spectra. The comparison of the calculated numbers with the numbers given by the supplier is presented in Table 5.1. The suppliers molar mass data have been determined by SEC.

As can be seen, a very good agreement between the MALDI-TOF and the SEC data is obtained. The gradually higher polydispersities calculated from the SEC measurements are attributed to the band broadening effects that are typical for SEC experiments.

It has already been described in Chap. 3 that the formation of molecular ions during the MALDI process can be promoted by the addition of cationizing agents. For different types of polymers different cations are most suitable, e.g., silver cations do promote molecular ion formation with polystyrene while alkali cations do not. On the other hand, alkali ions do promote the cationization of polyalkylene oxides. The effect of different alkali cations on the MALDI-TOF measurement of PEO is investigated in the next experiment. In this experiment, a PEO calibration standard with an average molar mass of roughly 4000 g/mol is measured using dithranol as the matrix and different alkali cations under otherwise similar conditions. The calculation of the molar mass averages is conducted as has been described previously and the results are summarized in Table 5.2.

The results indicate that under the present experimental conditions the different polarizabilities and the atomic radii of the cati-

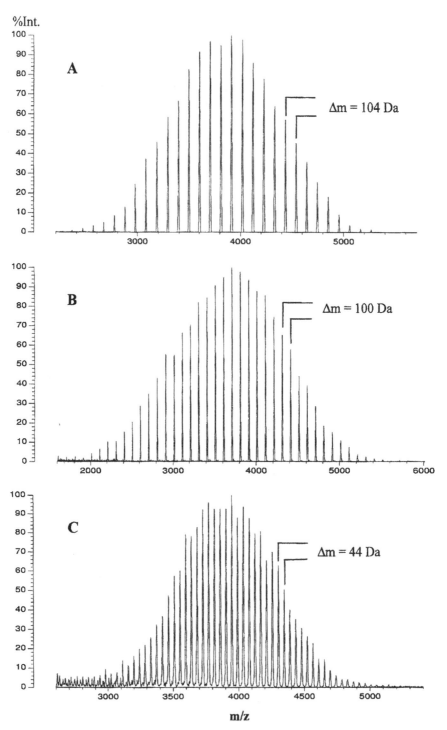

Fig. 5.5 A–C. MALDI-TOF spectra of narrowly disperse: **A** polystyrene; **B** PMMA; and **C** PEO, matrix: dithranol, linear mode

Table 5.1. Molar mass averages and polydispersities of polymers determined by MALDI-TOF and SEC (suppliers values)

Sample	SEC (supplier)			MALDI-TOF		
	M_w	M_n	M_w/M_n	M_w	M_n	M_w/M_n
PS	3700	3470	1.07	3820	3760	1.02
PMMA	3600	3250	1.10	3800	3700	1.03
PEO	4000	3700	1.07	3780	3540	1.07

Table 5.2. Molar mass analysis of PEO 4000 by MALDI-TOF, matrix: dithranol, cationizing agents: alkali metal trifluoroacetates

Cation	M_w	M_n	M_w/M_n
Li^+	3975	3955	1.005
Na^+	4045	4020	1.006
K^+	4060	4035	1.006
Rb^+	4045	4020	1.006
Cs^+	4050	4030	1.006

ons do not affect the mass spectrometric behavior of the sample. This might not be the case for other samples as has been shown in Fig. 5.1. It is, therefore, important to optimize not only the matrix but also the cationizing agent in order to obtain correct molar mass data.

5.2.2 Hydrocarbon Polymers [12]

While for water-soluble and polar organic polymers matrices can be used that have been applied originally for biopolymers, these matrices cannot be used for the analysis of non-polar polymers that are only soluble in organic solvents. For these types of polymers specific matrices must be applied that in terms of solubility and compatibility match the properties of the corresponding polymers. In particular, the analysis of hydrocarbon polymers like polystyrene, polybutadiene, and polyisoprene requires special attention due to the fact that these polymers do not ionize easily and particular efforts have to be devoted to the proper selection of the matrix and the cationizing agent.

Aim

The samples under investigation were commercial narrow disperse calibration standards for SEC. The polystyrenes were purchased from Polymer Laboratories (Churchstretton, GB) while the polybutadiene and the *cis*-1,4-polyisoprene were from Polysciences (Warrington, PA, U.S.A.).

Materials

Equipment MALDI-TOF system — Bruker Reflex MALDI-TOF, acceleration voltage of positive 30 kV, detection in linear or reflectron mode, 100–200 laser shots were summed per full spectrum.

MALDI-TOF sample preparation — Two methods for sample preparation have been used. Method 1: an equimolar mixture of IAA and silver acetylacetonate, Ag(acac), in methanol was subjected to 30 min of ultrasonication, filtered, and the filtrate dried. Varying ratios of the mixture and the pure IAA from 10:1 to 1:10 were then used as the matrix in THF and a concentration of 0.2 molar. This matrix solution was mixed in a 5:1 v/v ratio with the polymer solution (roughly 10^{-4} molar in THF) and 1–2 µl were applied to the probe and dried under ambient conditions. Method 2: saturated solutions of Ag(acac) and 1,4-di[2-(5-phenyl-oxazolyl)]benzene (POPOP) in THF were prepared by ultrasonication for 30 min. A mixture 10:1:1 v/v/v of the POPOP solution, the Ag(acac) solution and the polymer solution, respectively, was prepared and 1–2 µl were applied to the sample tip.

MALDI-TOF analysis

The MALDI-TOF spectrum of a polystyrene with a peak maximum molar mass of 5050 g/mol (PS 5k) is shown in Fig. 5.6. Sample preparation method 2 was used, detection was in the reflectron mode. The maximum of the distribution is observed at m/z 5166 Da and the distribution extends from 3000 to 7500 Da. The mass peaks are well separated and correspond to oligomers with a butyl endgroup. The sample has been prepared by anionic polymerization using tert-butyl lithium as the initiator. For a degree of polymerization of 47 the calculated oligomer mass is 4953 Da. As-

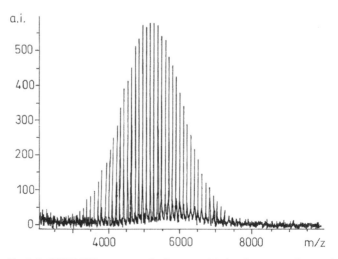

Fig. 5.6. MALDI-TOF spectrum of polystyrene PS 5k, reflectron mode, sample preparation method 2. (Reprinted from [12] with permission of Wiley, U.S.A.)

suming that ionization takes place via the attachment of a silver cation, this oligomer mass corresponds to an $[M+Ag]^+$ value of 5061 Da that is found in the spectrum. Accordingly, the next peak in the spectrum at 5166 Da corresponds to a degree of polymerization of 48.

The spectrum of polystyrene PS 11k is presented in Fig. 5.7. In this case sample preparation method 1 is used. As is shown here, there is a significant difference in resolution between the reflectron and the linear mode. For the reflectron measurement, baseline resolution occurs throughout the distribution and signal-to-

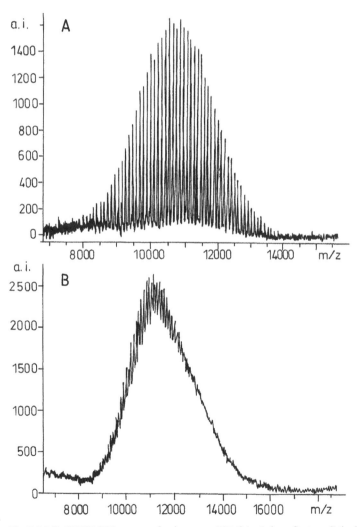

Fig. 5.7 A, B. MALDI-TOF spectra of polystyrene PS 11k in: **A** the reflectron; **B** the linear mode, sample preparation method 1. (Reprinted from [12] with permission of Wiley, U.S.A.)

noise ratio is very good. However, the maximum of the distribution appears at about 10,700 Da as opposed to the expected value from SEC of 11,600 Da. By comparison, the distribution measured in the linear mode is centered around 11,200 Da which is much closer to the expected value. This effect is explained by the channel depletion on the dual microchannelplate detector. When going to higher masses, one needs to collect more ions in order to obtain a similarly good signal-to-noise ratio. This is a particular problem as the distributions cover a larger m/z range. Since the integrated signal increases, one can reach the strip-current limit of the microchannelplate, and signal output for the later arriving ions is then lower than it should be. Therefore, for polymers with molar masses above approximately 10,000 g/mol these are determined more accurately using the linear detector.

The spectra of polystyrene with an M_p of 22,000 Da (PS 22k) using sample preparation methods 1 and 2 are shown in Fig. 5.8. Both spectra yield values of average molar masses that agree well with the expected values; see Table 5.3. The obvious difference between the spectra is the large extent of peaks observed at higher m/z in the case of sample preparation method 1. The peaks are the result of the aggregation of multiple polymer chains, forming clusters of up to ten polymer chains with m/z of 210,000. With method 2 much less clustering is obtained extending only to four to five polymer chains. The enhanced clustering with POPOP as the matrix is attributed to an agglomeration of the polymer chains in the solid matrix preparation and not due to gas phase clustering.

Table 5.3. Molar mass averages and polydispersities of polymers determined by MALDI-TOF and SEC (vendors values)

Sample	SEC (vendor)			MALDI-TOF		
	M_p	M_n	M_w/M_n	M_p	M_n	M_w/M_n
PS 5k	5,050	4,755	1.05	5,170	5,317	1.026
PS 11k	11,600	11,360	1.03	10,700[a]	10,560[a]	1.012[a]
				11,200[b]	11,520[b]	1.016[b]
PS 22k	22,000	21,410	1.03	21,100[c]	21,140[c]	1.009[c]
				21,000[d]	20,970[d]	1.010[d]
PS 125k	125,000		1.03	124,400	124,300	1.003
PB 10k	10,000		1.03	9,600	9,546	1.010
PI 11k	11,000		1.03	9,800	9,763	1.010

[a] Reflectron detector
[b] Linear detector
[c] Method 1
[d] Method 2

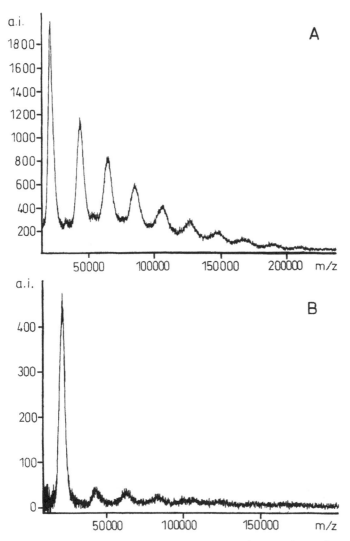

Fig. 5.8 A, B. MALDI-TOF spectra of polystyrene PS 22k, linear mode, sample preparation method: A 1; B 2. (Reprinted from [12] with permission of Wiley, U.S.A.)

While the styrene polymers yield relatively good signals with both sample preparation methods, the ions produced from polybutadiene and polyisoprene are of very low intensity. Only with method 2 and detection in the linear mode useful spectra are obtained; see Fig. 5.9.

In both cases the agreement with expected values of average molar masses is reasonable; see Table 5.3. No clustering is observed in either spectrum. Unfortunately, the ion signals are not intense enough for reflectron detection to allow one to confirm silver cationization, but it is assumed to occur based on results

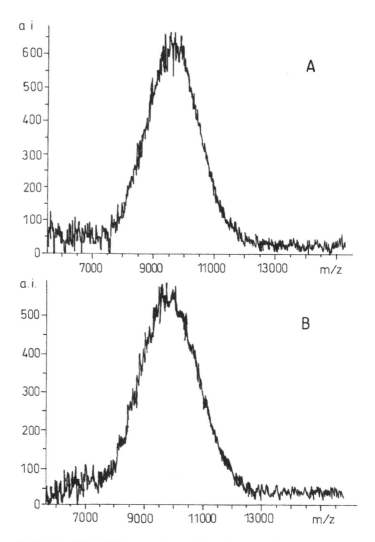

Fig. 5.9A,B. MALDI-TOF spectra of: **A** polybutadiene PB 10k; **B** polyisoprene PI 11k, linear mode, sample preparation method 2. (Reprinted from [12] with permission of Wiley, U.S.A.)

from SIMS [13,14] and laser desorption [15,16] which show good interaction between silver cations and unsaturated polymers.

Due to the higher polarizability of styrene, it is also reasonable that the interaction between silver cations and polystyrene is stronger than for polybutadiene and polyisoprene, and that this is the reason for the difficulty of ionization in the latter cases.

In conclusion, the sample preparation methods described here permit the MALDI-TOF analysis of PS, PB, and PI homopolymers. They should permit the analysis of a number of commercially important materials like ABS (acrylonitrile-butadiene-styrene copolymer), SBR (styrene-butadiene copolymer), and SAN (styrene-acrylonitrile copolymer).

5.2.3 Polymers with Varying Polydispersity [5]

As was discussed before qualitatively, the accuracy of molar mass measurements by MALDI-TOF mass spectrometry is a function of the polydispersity of the sample under investigation. In the following study polymers with varying polydispersity but similar chemical composition were prepared and analyzed. By comparing the MALDI results with results from SEC the effect of the polydispersity is evaluated.

Aim

PEG, PMMA, and polystyrene samples were purchased from Lab Service Analytica (Milano, Italy). They were prepared by anionic polymerization and analysed by SEC. The broadly distributed samples were prepared by free radical polymerization.

Materials

MALDI-TOF system — Bruker Reflex MALDI-TOF, acceleration voltage of positive 30kV, detection in linear mode, 150 laser shots were summed per full spectrum

Equipment

MALDI-TOF sample preparation — HABA was used as the matrix. The solvent was THF. Probe tips were loaded with 0.1 nmol polymer sample and 0.3 µmol matrix.

The MALDI-TOF spectra of the samples were taken using HABA as the matrix. For samples of lower molar mass where single oligomer peaks are resolved the peak height was taken as the molar fraction of that oligomer within the sample. For samples with no oligomer resolution, slicing procedures were used similar to conventional SEC quantification procedures.

MALDI-TOF analysis

The results of the MALDI-TOF and SEC measurements are summarized in Table 5.4. In both cases the most probable molar masses (M_p) and the polydispersities are given. $\Delta\%$ represents the difference between M_p (SEC) and M_p (MALDI).

Comparing the results for PMMA samples of different polydispersity it can be seen that the agreement between SEC and MALDI-TOF is very good for samples with very low polydispersity (1.04–1.06). In these cases the differences between SEC and MALDI-TOF lie in the range of 2.1–11.8%. These differences increase to 10.6–20.5% when the polydispersity increases to 1.08–1.10. As could be expected, for the high polydispersity sample PMMA-W1 the MALDI-TOF data are completely out of range. Instead of an expected molar mass of about 33,000 g/mol (from SEC) in MALDI-TOF only the low molar mass fraction is detected properly giving an M_p value of 2200 g/mol.

Table 5.4. Molar masses and polydispersities of polymers determined by MALDI-TOF and SEC

Sample	M_p (SEC)	M_p (MALDI)	Δ%	M_w/M_n (SEC)	M_w/M_n (MALDI)
PMMA 2400	2,400	2,100	12.5	1.08	1.10
PMMA 3100	3,100	2,700	12.9	1.09	1.11
PMMA 4700	4,700	4,200	10.6	1.10	1.08
PMMA 6540	6,540	5,200	20.5	1.09	1.11
PMMA 9400	9,400	7,500	20.2	1.10	1.08
PMMA 12700	12,700	10,400	18.1	1.08	1.10
PMMA 17000	17,000	15,000	11.8	1.06	1.03
PMMA 29400	29,400	27,000	8.2	1.06	1.01
PMMA 48600	48,600	47,000	2.1	1.05	1.01
PMMA 95000	95,000	90,000	5.3	1.04	1.01
PMMA-W1	33,000	2,200	94.0	2.50	1.15
PS 5050	5,050	5,100	1.0	1.05	1.02
PS 7000	7,000	7,020	0.3	1.04	1.02
PS 9680	9,680	9,600	0.8	1.02	1.01
PS 11600	11,600	11,300	2.6	1.03	1.02
PS 22000	22,000	20,800	5.5	1.03	1.01
PS 30300	30,300	28,000	7.6	1.03	1.02
PS 52000	52,000	46,000	11.5	1.03	1.01
PS-W2	9,000	2,000	77.0	2.00	1.06
PEG 4100	4,100	3,900	4.9	1.05	1.01
PEG 7100	7,100	7,420	4.3	1.03	1.02
PEG 8650	8,650	8,610	0.5	1.03	1.02
PEG 12600	12,600	12,790	1.5	1.04	1.01
PEG 23600	23,600	23,710	0.1	1.06	1.02

Similar results are obtained for PS and PEG. As long as the polydispersity is low, good agreeement between SEC and MALDI-TOF data is obtained. In a second series of experiments, polycondensation polymers like polyamide-6, polycarbonate, and polyesters are measured by SEC and MALDI-TOF. The polydispersities for these samples are in the range of 1.4–3.0 determined by SEC. For these samples MALDI-TOF yields M_p values that are 48–73% lower than the SEC values.

To summarize, the investigations clearly show that good agreement between SEC and MALDI-TOF experiments can be obtained only when the polydispersity of the samples is sufficiently low. For samples with higher polydispersity MALDI-TOF detects only the lower molar mass fraction of the distribution.

A graphical presentation of the effect of the polydispersity on the difference between MALDI-TOF and SEC results is given in

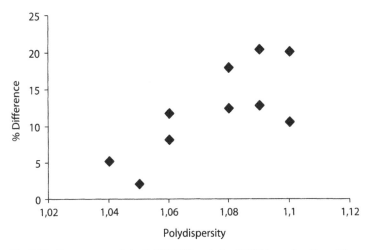

Fig. 5.10. Percent error of the MALDI-TOF analysis of PMMA as a function of the polydispersity of the sample

Fig. 5.10. It clearly shows that the percent error increases with increasing polydispersity.

5.3 Analysis of Polymers with Broad Molar Mass Distribution

It has been indicated in the previous section that for polydisperse polymers MALDI-TOF fails to produce reliable results. In many cases this problem can be solved by pre-fractionating the polydisperse sample using SEC. The resulting fractions with narrow molar mass distribution can then be measured accurately by MALDI-TOF. This approach will be addressed in detail in Chap. 7. In this section factors shall be discussed that cause errors in the measurement of broad polymer distributions. One major precondition for the accurate determination of molar masses is the detection of each molecule with equal efficiency. The investigation of this problem can be conducted by simulating broad polymer distributions through mixing of samples with narrow distribution but different molar masses.

5.3.1 Mixtures of Polyethylene Glycols

Narrow disperse polyethylene glycols are typical calibration standards for SEC. Their MALDI-TOF analysis is simple and good agreement between SEC and MALDI-TOF data is obtained; see

Aim

previous section. Due to their good ionizability in the low and high molar mass range, PEGs are useful model compounds for the simulation of broad molar mass distributions. In the following experiments the MALDI-TOF behavior of a binary mixture of PEGs with different molar masses shall be investigated.

Materials

The PEG samples under investigation were commercial narrow disperse calibration standards with molar masses of 1450 g/mol and 4000 g/mol for SEC. They were purchased from Polymer Standards Service GmbH (Mainz, Germany).

Equipment

MALDI-TOF system — Kratos Kompact MALDI III, acceleration voltage of positive 20 kV, detection in linear mode, 100–200 laser shots were summed per full spectrum

MALDI-TOF sample preparation — A matrix solution was prepared by dissolving dithranol in THF (10 mg/mL). The sample under investigation was dissolved in THF (3–5 mg/mL) and 10 μl of the sample solution were mixed with 10 μl of the matrix solution. To the resulting mixture another 1 μl of a 0.1 mol solution of the cationizing salt sodium trifluoroacetate was added. Then 0.5–1 μl of the final solution of sample, matrix, and salt were deposited on the sample target and air-dried.

MALDI-TOF Analysis

The following experiments are conducted with equimolar mixtures of two PEGs with molar masses of 1450 g/mol and 4000 g/mol. The polydispersity of each sample is lower than 1.1. As a first influencing factor, the effect of the laser power is investigated. In MALDI-TOF measurements the laser power required for the desorption/ionization process has to be adjusted slightly above the ionization threshold. This is only possible, however, in the case of monodisperse samples. When measuring mixtures, there is no well defined value for a threshold.

The behavior of the PEG mixture at two different laser powers is shown in Fig. 5.11. Taking the total peak area of each fraction as the measure for molar concentration, an area ratio of 1:1 for an equimolar mixture would indicate similar detection efficiencies. This is indeed the case at threshold intensity in the linear detection mode; see Fig. 5.11A. Even a slight increase in laser power such as 10% changes the picture. As can be seen in Fig. 5.11B for the corresponding measurement, the total peak area for the higher molar mass fraction is significantly higher. This indicates that higher molar masses require higher laser power than lower molar masses. In addition to laser power the stability of the produced ions is an influencing factor for the detection efficiency. As is shown in Fig. 5.11C, the higher molar mass fraction is detected with a much lower efficiency when a reflectron detector is used instead of a linear detector. Reflectron detection requires higher ion stability which seems not to be the case for the higher molar mass

5.3 Analysis of Polymers with Broad Molar Mass Distribution

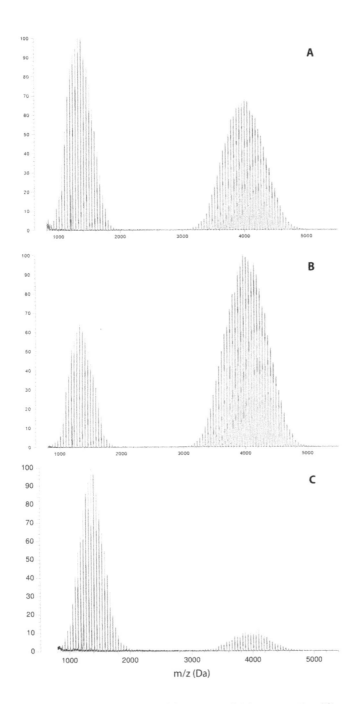

Fig. 5.11 A–C. MALDI-TOF spectra of the mixture of PEG 1450+4000 at different laser powers: **A** at threshold intensity; **B** at a laser power 10% above threshold in linear mode; **C** at threshold intensity in reflectron mode

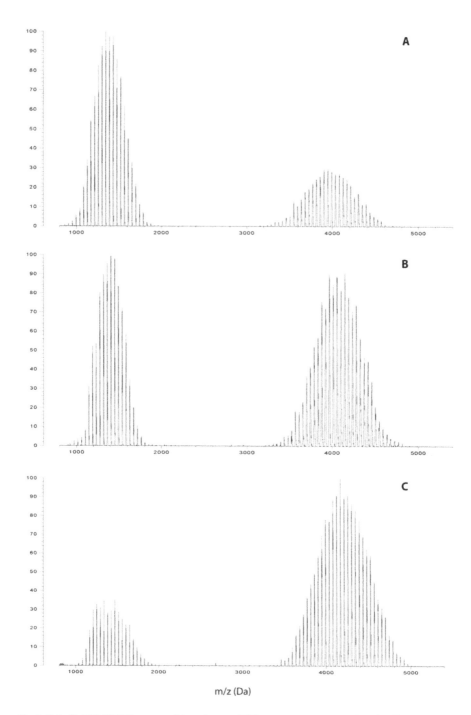

Fig. 5.12A-C. MALDI-TOF spectra of the mixture of PEG 1450+4000 cationized with different salts: **A** LiCl; **B** NaCl; **C** CsCl, linear mode

fraction. The importance of the laser power has already been emphasized in Sect. 3.4.5.

Another influencing factor is the type of cation that is attached to the polymer molecules; see Fig. 5.12. Under otherwise similar experimental conditions MALDI-TOF spectra of the PEG mixture are taken where Li^+, Na^+, or Cs^+ are used as the counterion. As is clear from the spectra, the detection efficiency of the higher molar mass fraction increases when the atomic radius of the counterion is increased.

5.3.2 Mixtures of Polystyrenes and Polymethyl Methacrylates [17]

Similar to the experiments decribed in Sect. 5.3.1 for polyethylene glycols, mixtures of polystyrenes and PMMAs are attractive model systems to mimic broad molar mass distributions. In the following experiments the effect of laser power is studied for mixtures, where the molar masses of the components exhibit a difference of one magnitude and more. — Aim

The PS and PMMA samples under investigation were commercial narrow disperse calibration standards for SEC. They were purchased from Polymer Standards Service GmbH (Mainz, Germany). — Materials

MALDI-TOF system — Bruker Reflex MALDI-TOF, acceleration voltage of positive 35 kV, detection in linear mode, 100–200 laser shots were summed per full spectrum. — Equipment

MALDI-TOF sample preparation — a matrix solution was prepared by dissolving dithranol in THF. The equimolar polymer mixtures PS (5500+20,800, 5500+46,000, 5500+98,700) and PMMA (6500+28,000, 6500+52,200, 6500+89,000) were dissolved in THF and added to the matrix at a molar ratio of approximately 1:0.02. PS and PMMA were cationized with silver trifluoroacetate and rubidium trifluoroacetate, respectively. The samples were crystallized on the MALDI target and measurements were carried out immediately.

Broad molar mass distributions are simulated by mixing well defined narrow distributed polymer standards. The low molar mass component is kept constant whereas the high molar mass component is increased from 20,000 to 100,000 g/mol to simulate increasing polydispersity. — MALDI-TOF Analysis

The effect of increasing laser power on the MALDI-TOF behavior of a mixture of PS 5500 and PS 46000 is shown in Fig. 5.13. At low laser power only the low molar mass distribution appears in the spectrum (Fig. 5.13A). As the laser power is increased, a small

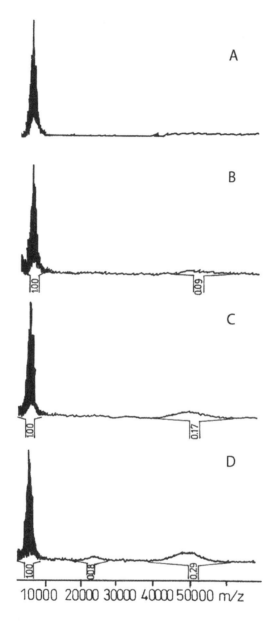

Fig. 5.13 A–D. MALDI-TOF spectra of the mixture of PS 5500+46000 at increasing laser power from A to D. (Reprinted from [17] with permission of Wiley, U.S.A.)

peak due to the high molar mass component appears (Fig. 5.13B) which further increases upon increasing the laser power (Fig. 5.13C,D). By increasing the laser power an optimal point can be reached, where the peak area does not grow further relative to

5.3 Analysis of Polymers with Broad Molar Mass Distribution

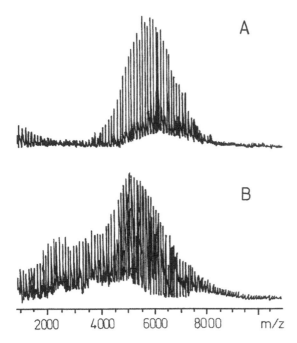

Fig. 5.14 A, B. MALDI-TOF spectra of the low molar mass component taken from Fig. 5.13; **A** low laser power; **B** high laser power. (Reprinted from [17] with permission of Wiley, U.S.A.)

the low molar mass distribution. If a laser power far beyond the optimal point is used, the low molar mass distribution becomes broader and is shifted to lower masses; see Fig. 5.14. Apparently, the high laser power causes fragmentation of the low molar mass molecules.

With increasing peak area of the high molar mass component the distribution of doubly charged ions at m/z 23,000 also increases; see Fig. 5.13. This does not disturb the analysis of narrow standards but in real broad distributions the doubly charged peaks may overlap with singly charged ions and will change the true peak intensity at this point.

Figure 5.15 shows the MALDI-TOF spectra of polystyrene mixtures with increasing molar mass of the higher molar mass component. In order to see the higher molar mass component a laser power must be used that causes partial fragmentation of the lower molar mass component; see Fig. 5.15 B, C. In addition, singly charged dimers and doubly charged molecules are observed which may disturb the true distributions.

Under the same conditions as described for polystyrene, mixtures of polymethyl methacrylate are prepared and analyzed. Fig-

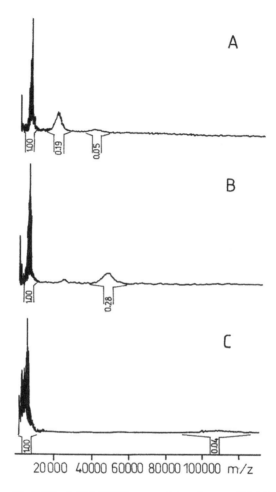

Fig. 5.15 A–C. MALDI-TOF spectra of the mixtures of PS: **A** 5500+20800; **B** 5500+46000; **C** 5500+98000. (Reprinted from [17] with permission of Wiley, U.S.A.)

ure 5.16 shows the corresponding MALDI-TOF spectra for mixtures with increasing molar masses of the higher molar mass component. Similar to PS, the high molar mass component can only be detected when a laser power is used that is well above the ionization threshold for the low molar mass component and causes partial fragmentation.

From the analysis of the polymer mixtures it can be concluded that the high molar mass part of a broad polymer distribution is always underestimated. A further problem comes from the different laser power required for analysis of low and high molar mass polymers of the same structure. This leads to different curve shapes of the obtained distributions. Thus, the average molar

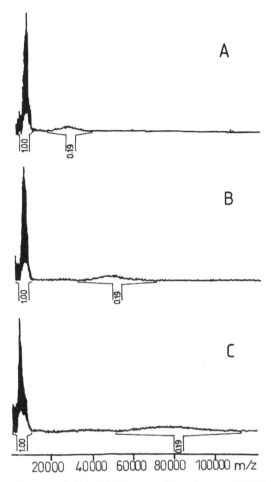

Fig. 5.16 A–C. MALDI-TOF spectra of the mixtures of PMMA: **A** 6500+28000; **B** 6500+52000; **C** 6500+89000; (Reprinted from [17] with permission of Wiley, U.S.A.)

mass values M_n and M_w will change depending on the laser power. For molar masses above 100,000 g/mol a partial fragmentation is obtained for the low molar mass part of the distribution. These fragments give additional signals which extend the distribution towards the low molar mass side. Finally, signals of doubly charged molecules and clusters also appear in the spectra.

Table 5.5. Applications of MALDI-TOF MS for molar mass determination

Polymer	Matrix	Remarks	Reference
Poly(dimethyl siloxane)	DHB	Analysis of PDMS by MALDI and SEC, prefractionation of PDMS, and MALDI analysis of the fractions	[18]
Poly(dimethyl siloxane)	DHB	Comparison of SEC, MALDI-TOF, TOF-SIMS, and ESI-FTMS	[19]
Poly(ethylmethyl siloxane)	Dithranol	Comparison of SEC, light scattering, viscometry, membrane osmometry and MALDI-TOF	[20]
Polystyrene	9-nitroan-thracene	Comparison of SEC and MALDI-TOF data	[6]
Polystyrene	Retinoic acid	NIST-sponsored interlaboratory comparison of a polystyrene of 7 kDa	[27]
PMMA	DHB	Comparison of SEC and MALDI-TOF data	[6]
PMMA	DHB	Determination of polymerization rate coefficients	[30]
PEG	DHB	Comparison of SEC, MALDI-TOF, TOF-SIMS, and ESI-FTMS	[19]
PEG	Dithranol	Determination of Mark-Houwink parameters	[31]
Poly(p-phenylene)	Dithranol	Analysis of low molar mass samples and comparison with HPLC data	[21]
PPG	THAP	Analysis of low molar mass samples and comparison with SFC, HPLC, and SEC data	[22]
Poly(tetrahydro pyrene)	Dithranol	Molar mass analysis of a rigid rod polymer, analysis of oligomers separated by SEC	[23]
Polycarbonate, poly(ethersulfone)	HABA	Analysis of samples with broad molar mass distribution	[24]
Polylactones	THAP, DHB	Analysis of macrocyclic polylactones and comparison with CI and FAB measurements	[25]
Epoxy oligomers	HABA	Delayed-extraction measurements	[26]

5.4 Further Applications

As has been pointed out earlier, attempts to measure polyolefins have been not very successful so far. Due to their inert nature and difficulty to dissolve in conventional solvents at room temperature they present special problems for sample preparation and ionization. In a recent work of Chen et al. [28] it has been shown that polyethylene samples of molar masses up to 4000 Da can be ionized by laser desorption/ionization TOF MS. It has been demonstrated that silver or copper ion attachment can occur in the gas phase. Fragmentation takes place under such conditions that can be reduced by adding a matrix to the sample preparation. In another study Bauer et al. [29] have presented evidence that it is possible to measure polyethylene after introducing polar groups by covalently bonding an organic cation to the polymer. To obtain ionizable molecules the polyethylene is dissolved in toluene and reacted with an excess of bromine. Afterwards triphenyl phosphine (TPP) is added to obtain a TPP-modified polyethylene. Using dithranol or all-*trans* retinoic acid as matrices, MALDI-TOF spectra could be produced in a mass range up to 7000 Da. The sample under investigation was a NIST standard reference material, SRM 2885.

References
1. DANIS PO, KARR DE, MAYER F, HOLLE A, WATSON CH (1992) Org Mass Spectrom 27:843
2. KARAS M, BAHR U, DEPPE A, STAHL B, HILLENKAMP F (1992) Macromol Symp 61:397
3. DANIS PO, KARR DE, WESTMORELAND DG, PITON MC, CHRISTIE DI, CLAY PA, KABLE SH, GILBERT RG (1993) Macromolecules 26:6684
4. SCHRIEMER DC, LI L (1996) Anal Chem 68:2721
5. MONTAUDO G, MONTAUDO MS, PUGLISI C, SAMPERI F (1995) Rapid Commun Mass Spectrom 9:453
6 LLOYD PM, SUDDABY KG, VARNEY JE, SCRIVENER E, DERRICK PJ, HADDLETON DM (1995) Eur Mass Spectrom 1:293
7 LLOYD PM, SCRIVENER E, MALONAY DR, HADDLETON DM, DERRICK PJ (1996) Polymer Prepr 37(1):847
8. BELU AM, DESIMONE JM, LINTON RW, LANGE GW, FRIEDMAN RM (1996) J Am Soc Mass Spectrom 7:11
9. BAHR U, DEPPE A, KARAS M, HILLENKAMP F, GIESSMANN U (1992) Anal Chem 64:2866
10. DANIS PO, KARR DE (1993) Org Mass Spectrom 28:923
11. JACKSON C, LARSEN B, MCEWEN C (1996) Anal Chem 68:1303

12. Danis PO, Karr DE (1996) Rapid Commun Mass Spectrom 10:862
13. Bletsos IV, Hercules DM, van Leyen D, Benninghoven A (1987) Macromolecules 20:407
14. Belu AM, Hunt MO, DeSimone JM, Linton RW (1994) Macromolecules 27:1905
15. Kahr MS, Wilkins CL (1993) J Am Soc Mass Spectrom 4:453
16. Mowat IA, Donovan RJ (1995) Rapid Commun Mass Spectrom 9:82
17. Martin K, Spickermann J, Räder HJ, Müllen K (1996) Rapid Commun Mass Spectrom 10:1471
18. Montaudo G, Montaudo MS, Puglisi C, Samperi F (1995) Rapid Commun Mass Spectrom 9:1158
19. Yan W, Ammon DM, Gardella JA, Marziarz EP, Hawkridge AM, Grobe GL, Wood TD (1998) Eur Mass Spectrom 4:467
20. Götz H, Maschke U, Wagner T, Rosenauer C, Martin K, Ritz S, Ewen B (2000) Macromol Chem Phys 201:1311
21. Remmers M, Müller B, Martin K, Räder HJ, Köhler W (1999) Macromolecules 32:1073
22. Trathnigg B, Maier B, Schulz G, Krüger RP, Just U (1996) Macromol Symp 110:231
23. Räder HJ, Spickermann J, Kreyenschmidt M, Müllen K (1996) Macromol Chem Phys 197:3285
24. Vitalini D, Mineo P, Scamporrino E (1997) Macromolecules 30:5285
25. Kricheldorf H, Eggerstedt S (1999) Macromol Chem Phys 200:1284
26. Schriemer DC, Whittal RM, Li L (1997) Macromolecules 30:1955
27. Guttman CM, Wetzel SJ, Blair WR, Fanconi BM, Girard JE, Goldschmidt RJ, Wallace WE, VanderHart DL (2001) Anal Chem 73:1252
28. Chen R, Yalcin T, Wallace WE, Guttman CM, Li L (2001) J Am Soc Mass Spectrom 12:1186
29. Bauer B, Wallace WE, Fanconi BM, Guttman CM (2001) Polymer 42:9949
30. Zammit MD, Davis TP, Haddleton DM (1996) Macromolecules 29:492
31. Tatro SR, Baker GR, Fleming R, Harmon JP (2002) Polymer 43:2329

6 Analysis of Complex Polymers

6.1 Introduction

The chemical structure of a complex polymer is characterized by its chemical composition, the sequence of the monomer units in the polymer chain, the functionality, and the molar mass. Typically, polymer analysis has to deal with the overlaying effects of different types of molecular heterogeneity in one macromolecular system. As for functional homopolymers (telechelics, macromonomers) the molar mass distribution is superimposed by a functionality type distribution. Random or segmented copolymers consist of a multitude of molecules with different chain lengths and different chemical compositions. In the ideal case, such complex systems should be separated into individual oligomers, which then may be analyzed with respect to chain lengths, functional groups, and chemical composition.

Resolution in liquid chromatography is not sufficiently high to separate complex polymer systems into individual oligomers in a broad molar mass range. Typically, oligomer resolution is achieved by liquid chromatography only up to molar masses of a few thousand g/mol. One of the most interesting features of MALDI-TOF mass spectrometry is that single polymer chain can be resolved to much higher molar masses. For conventional instruments with a continuous ion extraction geometry the upper limit for resolving single oligomer molecules is in the range of 10,000 to 15,000 Da. Newer developments such as delayed ion extraction (DE) improved the resolution of TOF instruments significantly, and with DE measurements oligomer separation can be achieved up to masses of 40,000 to 50,000 Da. With regard to mass resolution, MALDI-TOF is superior to all other techniques in analyzing chemically heterogeneous polymers with respect to the degree of polymerization, types of functional groups, and chemical composition of single polymer chains [1,2].

6.2 Determination of Endgroups

Oligomers and polymers with reactive functional groups have been used extensively to prepare a great variety of polymeric materials. In many cases, the behavior and reactivity of these functional homopolymers is largely dependent on the nature and the number of functional groups. In a number of important applications the functional groups are located at the end of the polymer chain. Macromolecules with terminal functional groups are usually termed "telechelics" or "macromonomers".

Molecular functionality, f, of a telechelic polymer is described as the number of functional groups per molecule. Macromolecules with the same structure of the polymer chain may be different in the *number* and the *nature* of the functional groups. Depending on the number of new bonds they can form in a reaction, one can classify functional groups as single-act groups, forming one new bond (-OH, -COOH, -NH$_2$, -COCl), and dual-act or multi-act groups, forming two or more new bonds (C=C, -CO-O-CO-, -N=C=O). To prepare a linear polymer by polymerization or polycondensation, each molecule must have f=2 for single-act or f=1 for dual-act groups. To obtain a cross-linked polymer, functionality must be f>2 or f>1, respectively.

When functional homopolymers are synthesized, functionally defective molecules are formed in addition to macromolecules of required functionality; see Fig. 6.1. For example, if a target func-

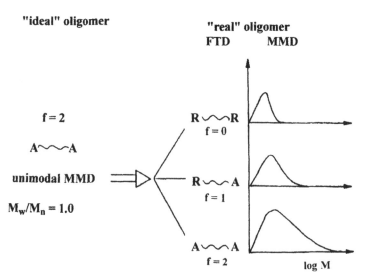

Fig. 6.1. Molar mass and functionality type distribution of a telechelic oligomer with A and R endgroups. (Reprinted from [3] with permission of Springer, Berlin Heidelberg New York)

6.2 Determination of Endgroups

tionality of f = 2 (A~A with two endgroups of type A) is required, then in the normal case species with f = 1 (R~A, only one endgroup of type A), f = 0 (R~R, no endgroup of type A) or higher functionalities are formed as well [3], which may result in a decreased or increased reactivity, cross-linking density, surface activity, etc. Each functionality fraction has its own molar mass distribution. Therefore, for complete description of the chemical structure of a functional homopolymer, the determination of the molar mass distribution and the functionality type distribution is required.

Typically, functionality is quantitatively described as a number-average functionality, f_n, where f_n is the ratio of the total number of functional groups to the total number of molecules in the system, i.e., the average number of functional groups per initial molecule. It is experimentally determined from

$$f_n = M_n / M_{eq} \tag{6.1}$$

where M_n is the number-average molar mass and M_{eq} is the equivalent molar mass, that is the average mass of the molecule per one functional group.

The f_n value provides information on the average functionality but does not characterize the functional polydispersity. An average functionality of 1 may be simulated by equal amounts of non-functional and two-functional species, and is therefore ambiguous. The characterization of the width of the functionality type distribution is more informative. In analogy to the average molar masses, number-average and weight-average functionalities may be introduced (see Sect. 1.1),

$$f_n = \Sigma n_i f_i / \Sigma n_i \tag{6.2}$$

$$f_w = \Sigma w_i f_i / \Sigma w_i = \Sigma n_i f_i^2 / \Sigma n_i f_i \tag{6.3}$$

where n_i is the number of molecules of functionality f_i, and $w_i = n_i f_i$.

For the description of the functional polydispersity the term f_w/f_n may be used. For polymers containing only one type of molecules, $f_w/f_n = 1$ is obtained; in the case of a distribution of molecules of different functionality $f_w/f_n > 1$ is obtained.

The following applications will describe different approaches to determine endgroups in a variety of polymers using MALDI-TOF mass spectrometry.

6.2.1 Polyamides [5]

Aim Polyamides are important technical polymers which are usually prepared by ring-opening condensation of lactames (e.g., polyamide-6) or co-condensation of a diamine and a diacid (e.g., polyamide-6.6). Typically, polymer chains with one amino and one carboxylic endgroup are formed. However, due to side reactions in technical processes or intentional endgroup modification reaction products are obtained which in addition to the molar mass distribution exhibit a functionality type distribution.

In the following application, polyamide-6 (PA-6) is modified to obtain products with different endgroups, including diamino-terminated, monoamino-terminated, and dicarboxy-termined species. The different types of functionality fractions are determined by MALDI-TOF.

Materials Commercial polyamide-6 (M_w=43,000 g/mol) is reacted with hexamethylene diamine to obtain diamino-terminated PA-6 (sample 1), with decylamine to obtain monoamino-terminated PA-6 (sample 2), with adipic acid to obtain dicarboxy-terminated PA-6 (sample 4), and with methanesulfonic acid to obtain lower molar mass α-amino-ω-carboxy-terminated PA-6 (sample 3). Some properties of the samples are summarized in Table 6.1. The endgroups are determined by titration.

Equipment MALDI-TOF system — Bruker Reflex MALDI-TOF, acceleration voltage of positive 30 kV, detection in reflectron mode, ions below m/z 350 were removed with pulsed deflection, 100 transients were summed per full spectrum

MALDI-TOF sample preparation — 2-(4-hydroxy phenylazo) benzoic acid (HABA) was used as the matrix and trifluoroethanol as the solvent. Probe tips were loaded with 100 pmol sample and 300 nmol matrix.

MALDI-TOF analysis For initial information on the MALDI-TOF behavior of the samples, the hydrolyzed sample 3 is analyzed. This sample is ex-

Table 6.1. Properties of the modified PA-6

Sample	[NH$_2$] (mmol/kg)	[COOH] (mmol/kg)	M_v[a] (g/mol)	M_n[b] (g/mol)
(1) NH$_2$ ~~~~~~ NH$_2$	572	---	7,000	3,000
(2) CH$_3$ ~~~~~~ NH$_2$	161	---	15,000	6,200
(3) HOOC ~~~~~~ NH$_2$	148	135	11,000	6,800
(4) HOOC ~~~~ COOH	19	312	19,000	6,400

[a] Determined by capillary viscometry
[b] Calculated from endgroup concentrations

6.2 Determination of Endgroups

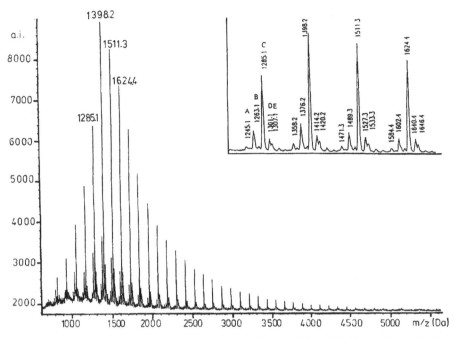

Fig. 6.2. MALDI-TOF spectrum of sample 3, matrix: HABA. (Reprinted from [5] with permission of Wiley, U.S.A)

pected to be α-amino-ω-carboxy-terminated. The spectrum of sample 3 in Fig. 6.2 is dominated by a series of intense peaks ranging from 800 to 5000 Da (series C), which can be assigned to the $[M+Na]^+$ molecular ions of α-amino-ω-carboxy-terminated PA-6. The four most intense peaks of this series, corresponding to oligomers with n = 12, 13, 14, and 15 are labeled in the spectrum and are reported enlarged in the inset of Fig. 6.2.

(A) $[-[-NH(CH_2)_5CO-]_n-]$ $[M+H]^+ = 1 + 113.1\,n$

(B) $H-[-NH(CH_2)_5CO-]_n-OH$ $[M+H]^+ = 19 + 113.1\,n$

(C) $H-[-NH(CH_2)_5CO-]_n-OH$ $[M+Na]^+ = 41 + 113.1\,n$

(D) $H-[-NH(CH_2)_5CO-]_n-OH$ $[M+K]^+ = 57 + 113.1\,n$

(E) $H-[-NH(CH_2)_5CO-]_n-ONa$ $[M+Na]^+ = 63 + 113.1\,n$

However, the spectrum also displays other mass peaks, which are obviously due to other PA-6 related oligomer series. Peak series B,

D, and E can be assigned to the same oligomer series, but ionized by the attachment of a proton (B), a potassium (D) or two sodia, respectively. Peak series A exhibits mass differences of 18 Da and 40 Da to series B and C, respectively, and can be assigned to protonated cyclic oligomers.

In a similar way, the MALDI-TOF spectra of samples 1 and 2 are taken; see Fig. 6.3. Sample 1 prepared by the reaction of PA-6 with hexamethylene diamine exhibits only one type of functionality fractions, namely the diamino-terminated oligomer series. The different series of mass peaks can be assigned to the $[M+H]^+$, $[M+Na]^+$, and $[M+K]^+$ molecular ions, series A, B, and C, respectively.

The spectrum of sample 2 exhibits cyclic oligomers in addition to the monoamino-terminated linear oligomers. Peak series (A) and (B) are assigned to the $[M+Na]^+$ and $[M+K]^+$ molecular ions of the cyclics, while series (C) and (D) correspond to the $[M+H]^+$ and $[M+Na]^+$ molecular ions of the following structure:

$$H\text{-}[\text{-}NH(CH_2)_5CO\text{-}]_n\text{-}NH\text{-}(CH_2)_9\text{-}CH_3$$

The reaction of the PA-6 with adipic acid results in a reaction which contains cyclic and α-amino-ω-carboxy-terminated oligomers in addition to the α,ω-dicarboxy-terminated PA-6 oligomers; see Fig. 6.4.

(1) $\overline{\text{-}[\text{-}NH(CH_2)_5CO\text{-}]_n\text{-}}$ $[M+H]^+ = 1 + 113.1\,n$

(2) $\overline{\text{-}[\text{-}NH(CH_2)_5CO\text{-}]_n\text{-}}$ $[M+Na]^+ = 23 + 113.1\,n$

(3) $HOOC\text{-}(CH_2)_4\text{-}CO\text{-}[\text{-}NH(CH_2)_5CO\text{-}]_n\text{-}OH$
    $[M+H]^+ = 147 + 113.1\,n$

(4) $H\text{-}[\text{-}NH(CH_2)_5CO\text{-}]_n\text{-}OH$ $[M+Na]^+ = 41 + 113.1\,n$

(5) $HOOC\text{-}(CH_2)_4\text{-}CO\text{-}[\text{-}NH(CH_2)_5CO\text{-}]_n\text{-}OH$
    $[M+Na]^+ = 169 + 113.1\,n$

(6) ???

(7) $HOOC\text{-}(CH_2)_4\text{-}CO\text{-}[\text{-}NH(CH_2)_5CO\text{-}]_n\text{-}ONa$
    $[M+Na]^+ = 191 + 113.1\,n$

6.2 Determination of Endgroups

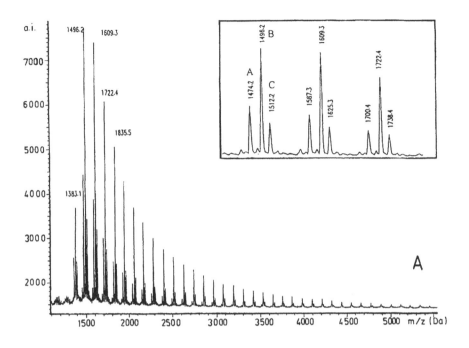

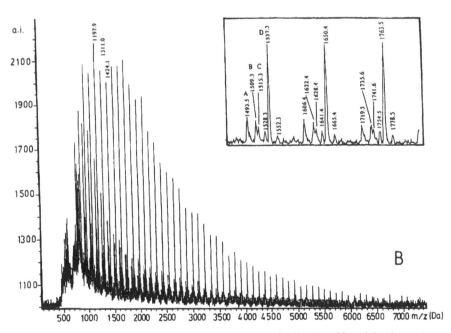

Fig. 6.3 A,B. MALDI-TOF spectra of samples: **A** 1; **B** 2, matrix: HABA. (Reprinted from [5] with permission of Wiley, U.S.A)

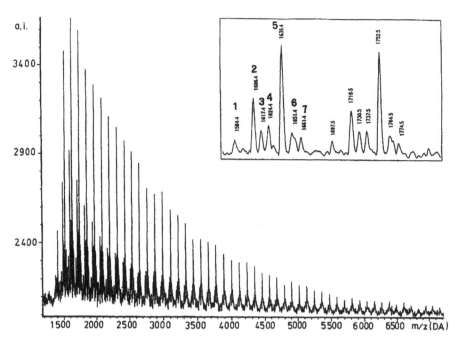

Fig. 6.4. MALDI-TOF spectrum of sample 4, matrix: HABA. (Reprinted from [5] with permission of Wiley, U.S.A)

The comparison of the MALDI-TOF spectra reveals that different from the other samples the spectrum for sample 1 does not show peaks for cyclic oligomers. A calculation of peak masses for linear and cyclic oligomers indicates that overlapping of mass peaks can be encountered. For example, the $[M+Na]^+$ molecular ion of the cyclic 13-mer has a mass of 1493.3 Da. The $[M+Na]^+$ molecular ion of the linear 12-mer (peak series B in Fig. 6.3A) has a mass of 1496.2 Da. The resolution of the TOF instrument in the present case was too low to separate these mass peaks. Therefore, the lower concentration cyclics cannot be detected in parallel to the higher concentration linears. To analyze the cyclic fraction, it must be extracted from the total sample and analyzed separately.

As has been shown, MALDI-TOF is a unique method for the determination of different functionality fractions in complex samples. Assuming that the peak intensities reflect the concentration profile for the different oligomers (which has to be verified), average functionalities f_n and f_w can be calculated from the oligomer peak areas; see Eqs. (6.2) and (6.3).

6.2.2 Triazine-based Polyamines [6]

Aromatic polyamides are important high temperature resistant polymeric materials. Their outstanding chemical stability and mechanical properties are accompanied, however, by difficulties in manufacture and processing. The synthesis of polymers, containing triazine rings instead of the aromatic dicarboxylic acid units in the polymer backbone by reaction of 6-substituted 2,4-dichloro-s-triazines with diamines results in triazine-based polyamines with interesting properties [7]. However, due to very poor solubility of the reaction products in organic solvents, the analysis of the chemical structure is very difficult. The reduced specific viscosity can be determined in formic and trifluoroacetic acids. SEC experiments, however, cannot be conducted due to a lack of suitable solvents. For the same reason a proper endgroup analysis cannot be carried out.

Aim

In MALDI-TOF experiments, formic acid is not a critical solvent. It can be used to dissolve the samples as well as the matrix 2,5-dihydroxy benzoic acid. Hence it should be possible to carry out an endgroup analysis of the polyamines by MALDI-TOF.

The synthesis of the triazine-based polyamines was carried out by interfacial polycondensation of 6-substituted 2,4-dichloro-s-triazines with various diamines [7]. The chemical structures of the components are given in Table 6.2.

Materials

Table 6.2. Triazine-based polyamines investigated by MALDI-TOF MS

Sample	Triazine	Diamine	Catalyst
1	O(CH₂CH₂)₂N—R (morpholine-N—R)	HN(CH₂CH₂)₂NH (piperazine)	NaOH
2	O(CH₂CH₂)₂N—R	HN(CH₂CH₂)₂NH	NaHCO$_3$
3	C_2H_5O-R	H_2N-$(CH_2)_6$-NH_2	NaOH
4	O(CH₂CH₂)₂N—R	H_2N—C$_6$H$_4$—NH_2	NaOH

R: 4,6-dichloro-1,3,5-triazin-2-yl

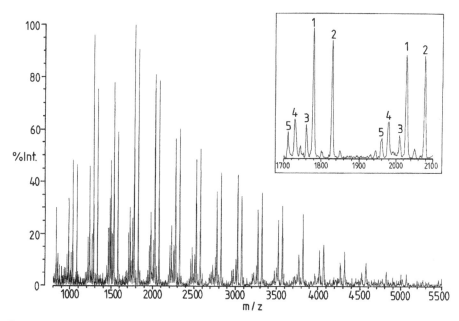

Fig. 6.5. MALDI-TOF spectrum of polyamine sample 1, matrix: DHB. (Reprinted from [6] with permission of Elsevier Science Ltd, UK)

Equipment

MALDI-TOF System — Kratos Kompact MALDI 3 with linear and reflectron detectors, acceleration voltage of positive 20 kV, detection in positive-reflectron mode, 100 transients were summed per full spectrum.

MALDI-TOF sample preparation — The samples were dissolved in formic acid in a concentration of 2–4 mg/ml and mixed with the matrix 2,5-dihydroxybenzoic acid (10 mg/ml solution in formic acid). After depositing 0.5 µl of the solution on the sample holder the solvent was evaporated in hot air.

MALDI-TOF analysis

Depending on the reactant ratios and the reaction conditions, oligomers having different endgroups can be formed. With an excess of the dichloro triazine in the reaction mixture, predominantly triazine endgroups are assumed to be formed. Vice versa, an excess of the diamine should lead predominantly to amine endgroups. With an equimolar ratio of the reactants, oligomers with mixed endgroups are likely to be formed.

The MALDI MS spectrum of a typical reaction product (sample 1) is shown in Fig. 6.5. Each peak in the spectrum represents one oligomer with a certain mass, characterizing the degree of polymerization (n) and the type of endgroups of this specific oligomer. The peaks can be assigned to different peak series having equal peak-to-peak mass increments of 249 Da. This mass increment ex-

6.2 Determination of Endgroups

actly equals the mass of the repeat unit in the polyamines. Accordingly, all peaks indicated with the same number belong to one homologous series.

At least five different homologous series are obtained in the spectrum, see numbers 1-5 in the inset of Fig. 6.5. They are due to intact oligomers cationized by the attachment of protons from the matrix to form [M+H]$^+$ molecular ions. The appearance of different homologous series having the same polymer backbone is obviously the result of the formation of different endgroups. From the total masses of the oligomers and the mass of the polymer backbone, the masses of the endgroups can be calculated and assigned to specific chemical structures. For example, taking the peak at 1779 Da and substracting the mass of the repeating unit several times, one ends up with a mass of 35 Da for the endgroups, assuming that the degree of polymerization n=7. Note that due to the low resolution of the measurement, chlorine isotopes cannot be detected:

$$[M+H]^+ = 1779 \text{ Da} = 36 + 249 \text{ n}$$

$$M = 1778 \text{ Da} = 35 + 249 \text{ n}$$

endgroups   polymer backbone

An endgroup mass of 35 Da can be assigned to an oligomer with the following chemical structure (accuracy of mass determination is ±1 Da):

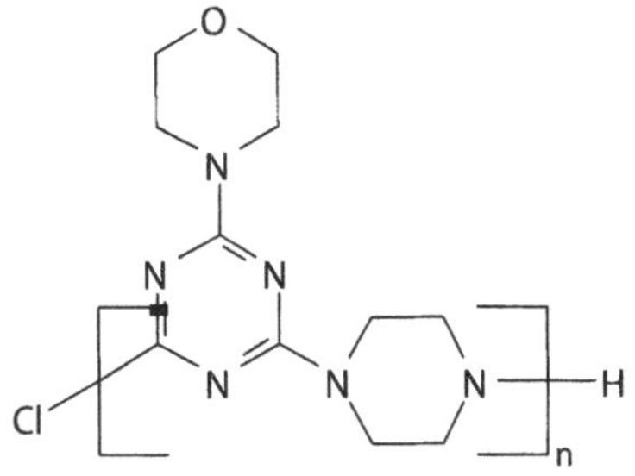

M (determined) = 1778 g/mol
M (calculated) = 1779 g/mol

In a similar way all other homologous series can be analyzed, giving the masses of the endgroups and their chemical structures, accordingly:

(1)   M =   35 + 249 n
(2)   M =   87 + 249 n
(3)   M =   17 + 249 n
(4)   M = 237 + 249 n
(5)   M = 217 + 249 n

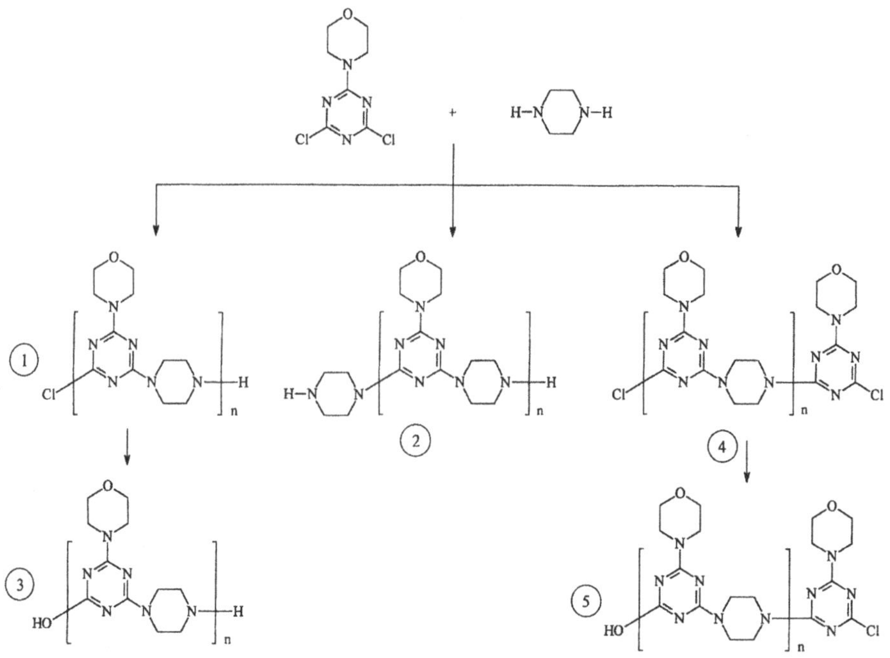

Fig. 6.6. Reaction mechanism of the polycondensation of 2,4-dichloro-6-morpholinyl-s-triazine and piperazine. (Reprinted from [6] with permission of Elsevier Science Ltd, UK)

In agreement with the reaction conditions, homologous series with an amino and a chlorotriazine endgroup (series 1) and two amino endgroups (series 2) can be identified as the main fractions. In addition, a third fraction with two chlorotriazine endgroups (series 4) is present in the reaction product. The homologous series 3 and 5 are assumed to be formed by partial hydrolysis of the chlorine at the triazine endgroup to a hydroxy function. The proposed reaction mechanism is summarized in Fig. 6.6.

The hydrolysis of the chlorotriazine endgroups is a side reaction which leads to endgroup deactivation, and chain termination, respectively. This reaction is obviously accelerated by the aqueous NaOH present in the reaction mixture. Assuming that a weaker base would reduce hydrolysis, NaHCO$_3$ was used instead of NaOH. The MALDI MS spectrum of the reaction product, shown in Fig. 6.7, clearly confirms this assumption. Peak series 3 and 5 which belong to the hydrolysis products are no longer present in the spectrum.

In another experiment, 6-ethoxy-2,4-dichloro-s-triazine is reacted with hexamethylene diamine (sample 3). In this case the MALDI MS spectrum indicates that only oligomers with triazine endgroups are formed; see Fig. 6.8. Similar to the previous results,

6.2 Determination of Endgroups

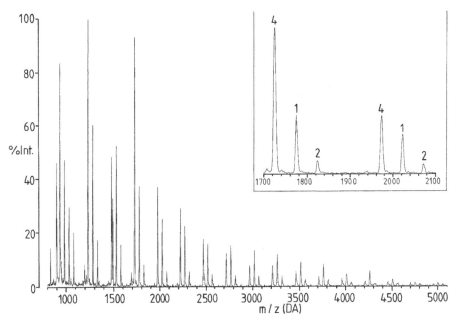

Fig. 6.7. MALDI-TOF spectrum of polyamine sample 2, matrix: DHB. (Reprinted from [6] with permission of Elsevier Science Ltd, UK)

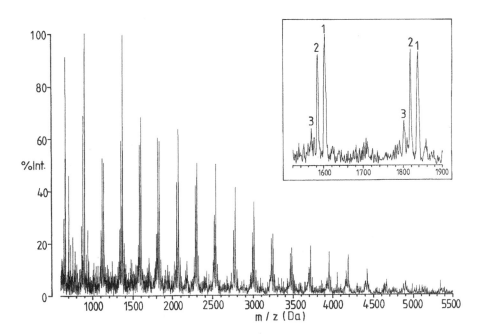

Fig. 6.8. MALDI-TOF spectrum of polyamine sample 3, matrix: DHB. (Reprinted from [6] with permission of Elsevier Science Ltd, UK)

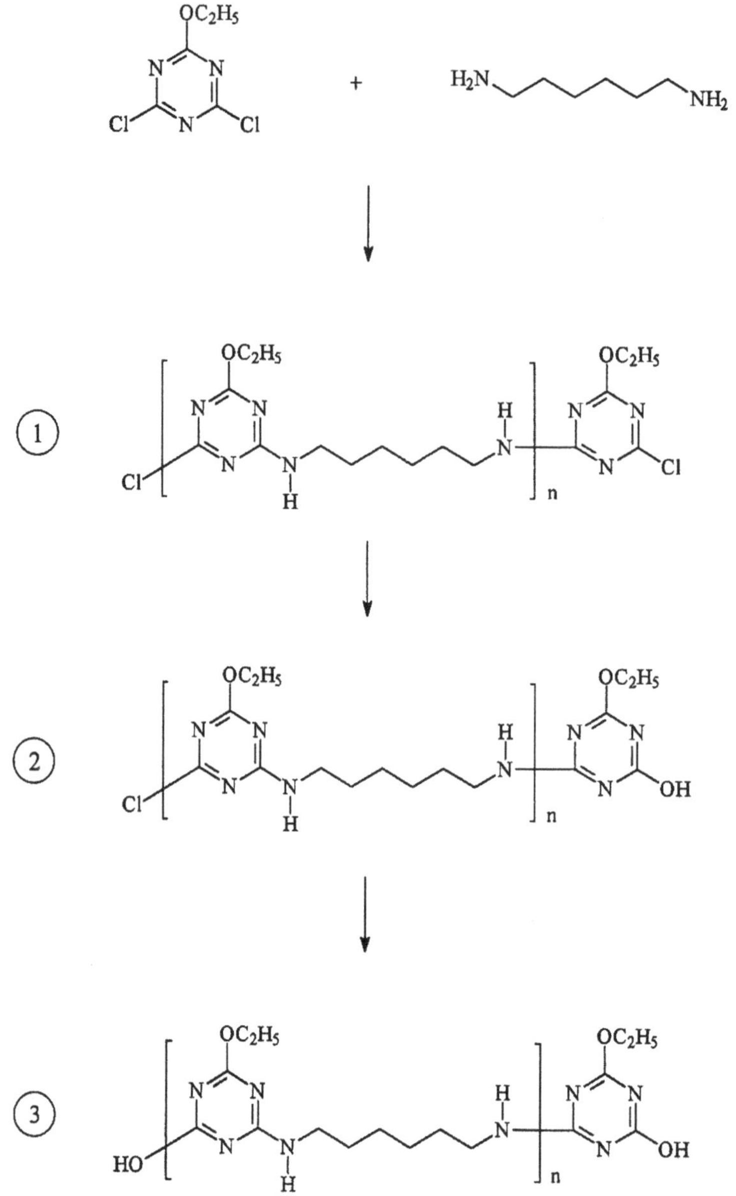

Fig. 6.9. Reaction mechanism of the polycondensation of 2,4-dichloro-6-ethoxy-s-triazine and 1,6-hexamethylene diamine. (Reprinted from [6] with permission of Elsevier Science Ltd, UK)

species with hydroxytriazine endgroups are formed to a significant extent, indicating that partial hydrolysis has to be accounted for; see reaction scheme in Fig. 6.9.

The reaction of 2,4-dichloro-6-morpholinyl-s-triazine with p-phenylene diamine (sample 4) leads to a colored product, which

6.2 Determination of Endgroups

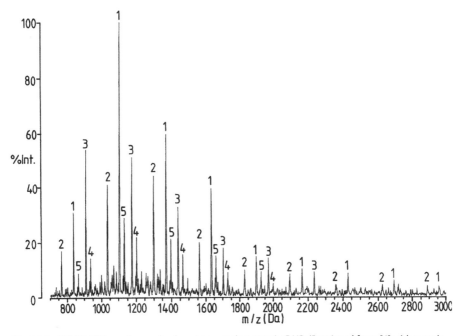

Fig. 6.10. MALDI-TOF spectrum of polyamine sample 4, matrix: DHB. (Reprinted from [6] with permission of Elsevier Science Ltd, UK)

even after extensive purification procedures kept its color. Therefore, it must be assumed that the color is not due to any impurity in the reaction mixture but due to the incorporation of a colored moiety into the polymer chain.

The MALDI MS spectrum of sample 4 shows that in addition to the expected homologous series 1, 2, and 3 two other homologous series (indicated with numbers 4 and 5) are obtained; see Fig. 6.10.

This colored diamine, of course, can also participate in the polycondensation reaction with the dichloro triazine. Accordingly, 2,7-diamino phenazine may be incorporated into the polymer chain or at the chain end. The calculation of the mass of the endgroup in series 4 agrees well with this assumption, and indeed series 4 is a homologous series having a phenazine endgroup; see Fig. 6.11. The calculation of the mass of the endgroup in series 5 gives 65 Da. Unfortunately, this mass could not be correlated to a specific chemical structure.

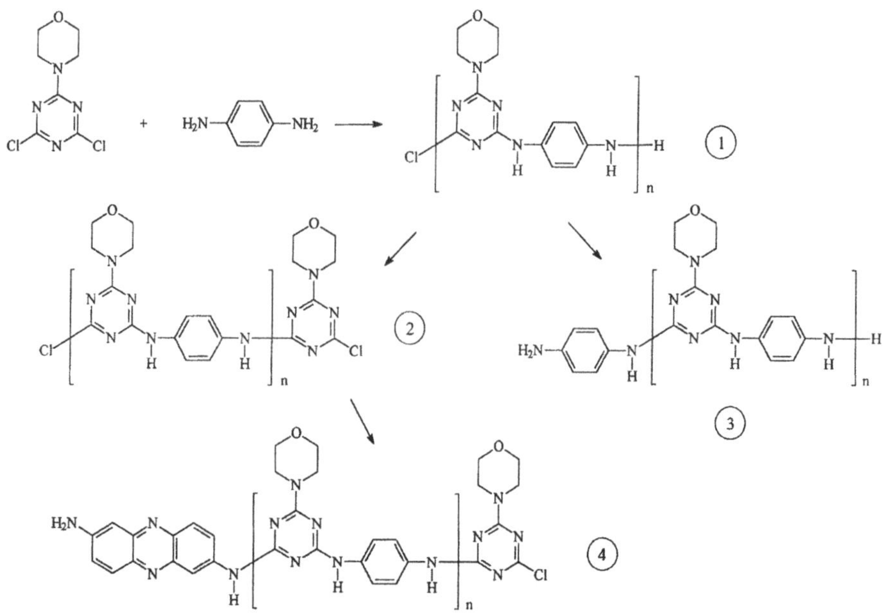

Fig. 6.11. Reaction mechanism of the polycondensation of 2,4-dichloro-6-morpholinyl-s-triazine and phenylene diamine. (Reprinted from [6] with permission of Elsevier Science Ltd, UK)

6.2.3 Poly(methylphenyl silane) [8]

Aim

In endgroup characterization by mass spectrometry, a difficulty often encountered is given by the possibility that two or more structures corresponding to different cationized oligomers show isobar masses. Therefore, the mass peak assignment is made ambiguous. To solve this ambiguity, the addition of different cations to the sample can be used. Since polymer molecules attach K^+, Na^+, or Li^+ to form molecular ions, the different affinity of molecules of different chemical structures towards different cations can be used to obtain unambiguous results. One may compare spectra obtained from a sample with and without addition of a specific cation, or after addition of different cations.

The systematic use of the technique of doping MALDI mass spectra with specific salts is applied in the present example to determine the endgroups in a sample of poly(methylphenyl silane). By using this approach, different types of linear and cyclic oligomer fractions can be detected in the sample.

Materials

The sample of poly(methylphenyl silane) was prepared by Wurtz coupling of the corresponding dichloro methylphenylsilane with sodium in toluene [9].

6.2 Determination of Endgroups

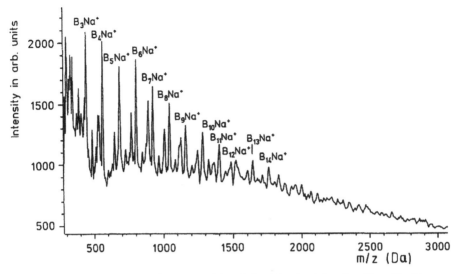

Fig. 6.12. MALDI-TOF spectrum of poly(methylphenyl silane) fraction 1. (Reprinted from [8] with permission of Wiley-VCH, Germany)

MALDI-TOF system — Bruker Reflex MALDI-TOF, acceleration voltage of positive 30kV, detection in reflectron mode, ions below m/z 350 were removed with pulsed deflection, 100 transients were summed per full spectrum

Equipment

MALDI-TOF sample preparation — 2,5-dihydroxybenzoic acid (DHB), dithranol and 3-indoleacrylic acid (IAA) were used as the matrix and THF as the solvent. Probe tips were loaded with 20 pmol of sample and 100 nmol of matrix.

The synthesis of polysilanes via Wurtz coupling is a heterogeneous polymerization reaction, which may produce polydisperse materials containing cyclic oligomers and open-chain oligomers with different endgroups. In order to get information on the distribution of the endgroups as a function of the molar mass, the sample is fractionated by precipitation in methanol. The resulting three fractions are analyzed by MALDI-TOF. Due to the high polydispersity of the sample, the spectra reflect only the low molar mass portion of the distribution.

MALDI-TOF analysis

The MALDI-TOF spectrum of fraction 1, recorded in the linear mode, is presented in Fig. 6.12. The spectrum shows well resolved signals that can be assigned to four distinct mass peak series A, B, C, and D; see the expanded portion of the spectrum in Fig. 6.13A. The mass peak series A–D can be assigned to the following oligomer series, where series A–C correspond to linear oligomers, while D is due to cyclic structures:

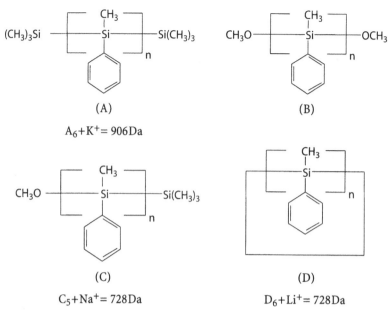

(A)

$A_6 + K^+ = 906\,Da$

(B)

(C)

$C_5 + Na^+ = 728\,Da$

(D)

$D_6 + Li^+ = 728\,Da$

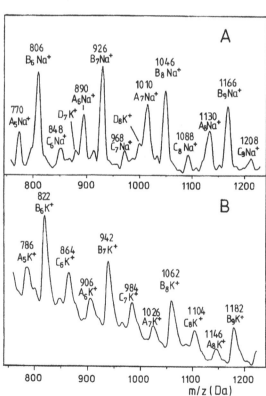

Fig. 6.13 A,B. Enlarged portions of the MALDI-TOF spectra of poly(methylphenyl silane) fraction 1: **A** matrix IAA; **B** matrix dithranol+KCl. (Reprinted from [8] with permission of Wiley-VCH, Germany)

6.2 Determination of Endgroups

Due to their natural abundance, different types of cations can be formed including the $[M+Li]^+$, $[M+Na]^+$, and $[M+K]^+$ molecular ions. A calculation of the corresponding mass numbers reveals that the peak series C_n+Na^+ is isobar to $D_{n+1}+Li^+$. To solve this ambiguity, a second MALDI-TOF experiment is carried out, doping the solution with KCl. This procedure should enhance the intensity of the $[M+K]^+$ mass peaks, and the $[M+Li]^+$ and $[M+Na]^+$ peaks should be shifted by 32 Da and 16 Da, respectively. In Fig. 6.13B the expanded portion of the corresponding spectrum is presented, showing the expected mass shifts of 16 Da for the mass series A, B, and C. This fact allows the assignment of the peak series C in Fig. 6.13A to the structure C_n+Na^+, since the alternative structure $D_{n+1}+Li^+$ would exhibit a mass shift of 32 Da.

In a similar way, fraction 2 of the poly(methylphenylsilane) is analyzed; see Figs. 6.14 and 6.15.

The expanded portion of the spectrum in Fig. 6.15A shows the presence of at least eight distinct mass series. These can be assigned to differently cationized oligomer series D, E, F, and G.

(D)
$D_6+K^+ = 760\,\text{Da}$

(E)
$E_6+Li^+ = 760\,\text{Da}$

(F)
$F_6+Na^+ = 906\,\text{Da}$

(G)

As in the previous case, the assignments of the mass series E_n+Li^+ and F_n+Na^+ might be ambiguous due to the isobar mass series D_n+K^+ and A_n+K^+, respectively. Therefore, doping with sodium iodide was used to confirm the assignments; see Fig. 6.15B. In this case, only $[M+Na]^+$ molecular ions are formed which can be identified readily.

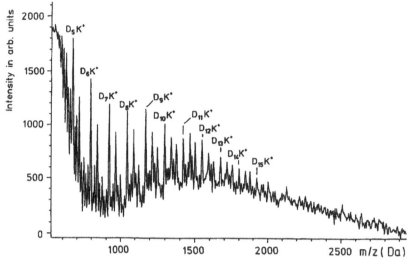

Fig. 6.14. MALDI-TOF spectrum of poly(methylphenyl silane) fraction 2. (Reprinted from [8] with permission of Wiley-VCH, Germany)

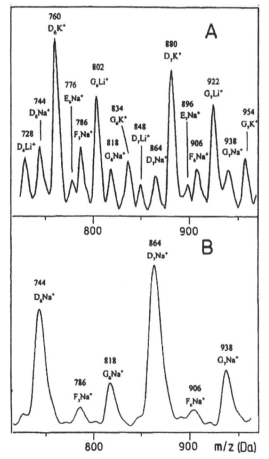

Fig. 6.15 A,B. Enlarged portions of the MALDI-TOF spectra of poly(methylphenyl silane) fraction 2: **A** matrix DHB; **B** matrix dithranol+NaI. (Reprinted from [8] with permission of Wiley-VCH, Germany)

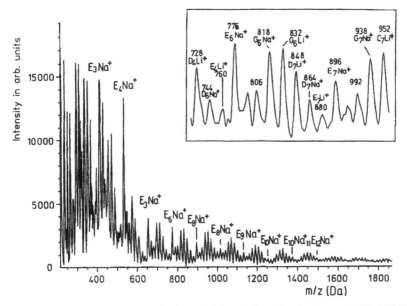

Fig. 6.16. MALDI-TOF spectrum of poly(methylphenyl silane) fraction 3, matrix DHB+LiCl. (Reprinted from [8] with permission of Wiley-VCH, Germany)

Finally, fraction 3 of the poly(methylphenyl silane) is analyzed by MALDI-TOF; see Fig. 6.16. In this case, from the very beginning LiCl is added to the matrix DHB to enhance the formation of the $[M+Li]^+$ molecular ions. As can be seen in the enlarged part of the spectrum, indeed strong mass peaks for the lithium-cationized oligomers are obtained. Although the formation of $[M+Na]^+$ molecular ions is not fully suppressed, an unambiguous assignment of the different peak series is possible. In this fraction, low molar mass cyclic oligomers (D) are present together with linear oligomers having different endgroups (C, E, G).

To summarize, MALDI-TOF shows clearly that different fractions of poly(methylphenyl silane) contain oligomers with different endgroups.

6.2.4 GTP Polymerized PMMA [10]

Group-transfer polymerization (GTP) is an interesting technique of living polymerization for polymethacrylates. In this technique, a silyl ketene acetal initiator reacts with a monomer by a Michael addition. During the addition, the silyl group transfers to the monomer, generating a new ketene acetal function. In this way, living polymethacrylates are formed at ambient temperatures.

Aim

GTP is attractive because polymethacrylates with a molar mass distribution M_w/M_n close to 1 are obtained.

Polymers with reactive functionality at one end are useful tools for the synthesis of graft copolymers and ABA block copolymers. To make sure that copolymers with strictly predetermined structures are formed, a functionality analysis of the precursor polymers is necessary. For the analysis of GTP-polymerized polymethyl methacrylates with respect to molar mass and functionality, MALDI-TOF shall be used. The functional heterogeneity obtained by MALDI-TOF shall be compared with data obtained by SEC.

Materials

The PMMA samples were prepared by GTP according to the following reaction scheme. They are typically used as calibration standards for SEC [11]. The nominal molar masses of the samples were 720 g/mol for sample 1, 900 g/mol for sample 2, 1800 g/mol for sample 3, 1320 g/mol for sample 4, 1420 g/mol for sample 5, and 1800 g/mol for sample 6.

Equipment

MALDI-TOF system — Kratos Kompact MALDI 3 with linear and reflectron detectors, acceleration voltage of positive 20 kV, detection in positive-reflectron mode, 100 transients were summed per full spectrum.

MALDI-TOF sample preparation — the samples were dissolved in THF (3 mg/mL) and mixed 1:1 with the matrix 2,5-dihydroxybenzoic acid (10 mg/mL solution in THF). For the enhancement of ion formation, a small amount of LiCl was added to the solution. After depositing 0.5 µl of the solution on the sample holder the solvent was evaporated in hot air.

MALDI-TOF analysis

According to the reaction scheme for GTP, macromolecules with two hydrogen atoms as endgroups are obtained. A heterogeneity in the endgroups is not expected. This is in perfect agreement with the MALDI-TOF spectrum of sample 1, given in Fig. 6.17A.

6.2 Determination of Endgroups

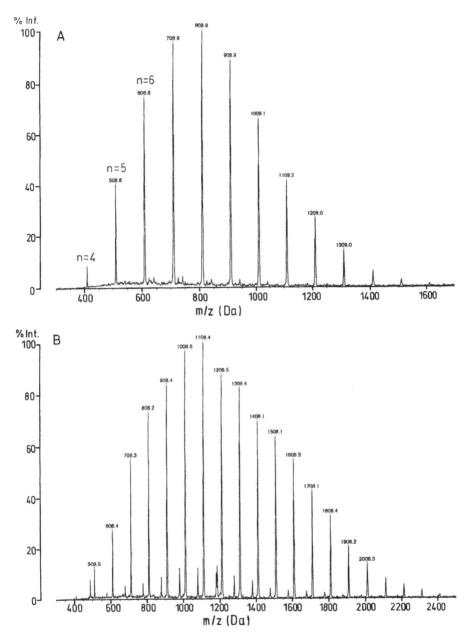

Fig. 6.17 A,B. MALDI-TOF spectra of GTP-polymerized samples: **A** sample 1; **B** sample 2, matrix DHB+LiCl. (Reprinted from [10] with permission of Elsevier Science Ltd, UK)

The mass peaks correspond to the $[M+Li]^+$ molecular ions. Their formation is enhanced by the addition of LiCl to the matrix. As expected, only one oligomer series is obtained corresponding to the proposed structure of the macromolecules ($M+Li^+ = 9+100n$, n being the degree of polymerization). As opposed to sample 1, the higher molar mass sample 2 in Fig. 6.17B exhibits a second peak series of lower intensity. The peak-to-peak mass increment of this series is 100 Da, indicating that it is a second PMMA-based homologous series. The mass shift between the two peak series is 31 Da. As the mass increment of both homologous series is the same, the changes in the chemical structure must be attributed to variations in the endgroups. Similar observations have been made for GTP-polymerized polybutyl methacrylate [12]. The fact that both series have their maximum abundance at about 1100 Da indicates that they are formed simultaneously in the reaction.

For a sample of higher molar mass (sample 3, average molar mass 1800 g/mol) the oligomer distributions of the peak series are quite different; see Fig.

6.18. While the abundance maximum of the major series is at 2100 Da, it is located at 1200 Da for the second series. This might indicate a molar mass effect in the formation of this series.

To investigate the nature of the second oligomer series (●) the low molar mass PMMA sample 5 is separated into oligomers by high-resolution SEC. The chromatogram in Fig. 6.19A indicates the presence of a peak (C_3) that does not fit into the oligomer series. This peak is separated preparatively and identified by NMR as being the cyclic trimer. In MALDI-TOF it appears as the $[M+Li]^+$ molecular ion at 278 Da. The same peak is detected in the

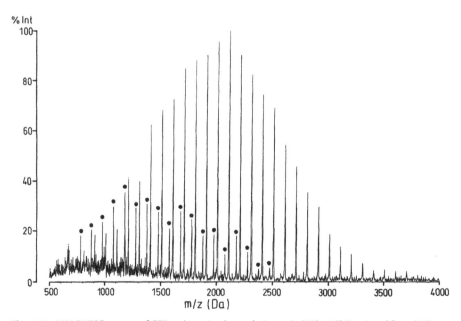

Fig. 6.18. MALDI-TOF spectra of GTP-polymerized sample 3, matrix DHB+LiCl. Reprinted from [10] with permission of Elsevier Science Ltd, UK)

6.2 Determination of Endgroups

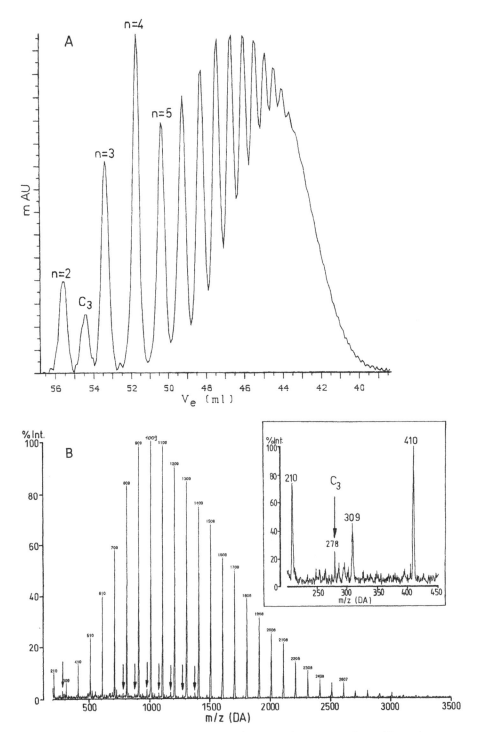

Fig. 6.19 A,B. SEC of GTP-polymerized sample 5 (**A**). MALDI-TOF spectrum of same (**B**), matrix DHB+LiCl. (Reprinted from [10] with permission of Elsevier Science Ltd, UK)

MALDI-TOF spectrum of the total sample, exhibiting a mass shift towards the linear trimer of 31 Da; see Fig. 6.19B.

Further peaks of low intensity appearing at 778, 878, 978 Da, etc. can be attributed to the second oligomer series (indicated by an arrow). It is obvious, that this series is due to the formation of oligomers with cyclic endgroups in addition to the linear oligomers. The formation of these endgroups is shown in the following reaction scheme:

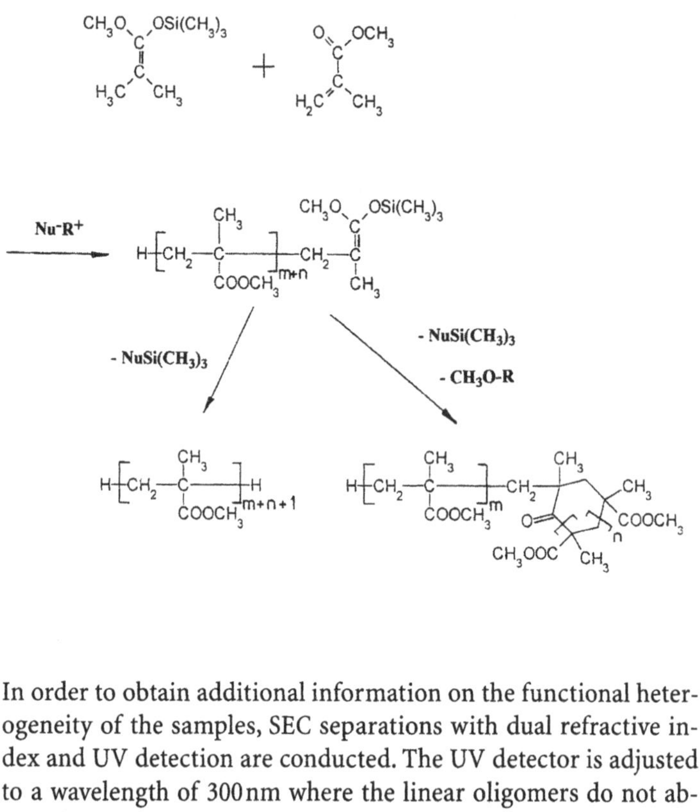

In order to obtain additional information on the functional heterogeneity of the samples, SEC separations with dual refractive index and UV detection are conducted. The UV detector is adjusted to a wavelength of 300 nm where the linear oligomers do not absorb light. Owing to the keto ester structure of the cyclic endgroup, these oligomers have an absorption maximum at about 300 nm. Therefore, the UV trace represents the molar mass distribution of the oligomers with cyclic endgroups whereas the RI trace characterizes the total sample; see Fig. 6.20 for sample 4. These molar mass distributions are compared to oligomer distributions calculated from the MALDI-TOF spectra; see Table 6.3. Considering the different problems associated with quantification from MALDI-TOF spectra, a very satisfying agreement between the different data is obtained.

6.2 Determination of Endgroups

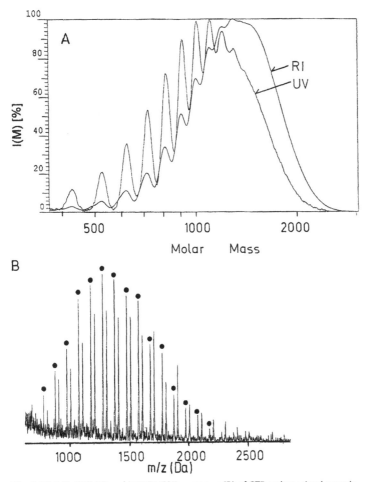

Fig. 6.20 A,B. SEC (**A**) and MALDI-TOF spectrum (**B**) of GTP-polymerized sample 4, matrix DHB+LiCl (Reprinted from Ref. [10] with permission of Elsevier Science Ltd, UK)

Table 6.3. Molar mass data of the PMMA samples from SEC and MALDI-TOF

Sample	M_n (SEC)	M_n (MALDI-TOF)
3 Total	1800	
Linear		1900
Cyclic		1310
4 Total	1220	
Linear		1450
Cyclic	1040	1190
6 Total	1690	
Linear		1850
Cyclic	1620	1740

6.2.5 Poly-ε-Caprolactone [13]

Aim

Covalent metal carboxylates, particularly tin(II)octoate (Sn(Oct)$_2$), belong to the most frequently used initiators for polymerization of cyclic esters. The most advocated mechanism is a direct catalytic action of Sn(Oct)$_2$ by activating the monomer, forming a donor-acceptor complex and participating directly in chain propagation [14,15]. It follows from this mechanism that Sn(II) atoms are not covalently bound to the polymer chain at any stage of polymerization. In another mechanism proposed Sn(Oct)$_2$ reacts with compounds containing hydroxy groups to form the actual initiator tin(II)alkoxide or tin(II)hydroxide [16–18]. The elementary reaction of the chain growth is assumed to proceed via monomer insertion and then tin(II) atoms are present at the polymer chain end. However, neither kinetic nor spectroscopic data have been available to support one of the two mechanisms.

In the present study, MALDI-TOF is used to detect Sn(II)-containing species incorporated at the endgroups in one of the populations of macromolecules. This is the first observation of the active species in cyclic ester polymerization by MALDI-TOF.

Materials

The polymerizations were carried out by mixing and heating ε-caprolactone, tin octoate, butanol or water, and THF in sealed glass ampoules. A detailed description is given in [13].

Equipment

MALDI-TOF system — Voyager-Elite of Perseptive Biosystems (Framingham, U.S.A.) with linear and reflectron detectors, acceleration voltage of positive 20 kV, detection in positive-reflectron mode with delayed extraction.

MALDI-TOF sample preparation — The samples were dissolved in THF (1 mol/l) and mixed 25:1 with the matrix 2,5-dihydroxybenzoic acid (10 mg/ml solution in THF). The mixture was dried on a stainless steel target covered by gold.

MALDI-TOF analysis

Successful observation of Sn(II)-containing species by MALDI-TOF requires two prerequisites to be fulfilled. First, the starting concentration of Sn(Oct)$_2$ has to be high enough, and second, ultimate care has to be taken in preparing and handling the polymer solutions to avoid hydrolysis. A high starting concentration of Sn(Oct)$_2$ is needed to have macromolecules with a sufficiently low molar mass to observe the characteristic Sn isotope fingerprint. This unique fingerprint results from the distribution of ten, naturally abundant, Sn isotopes which are present in comparable concentrations.

Figure 6.21 shows the MALDI-TOF spectrum of a polycaprolactone (PCL) prepared with a Sn(Oct)$_2$/butanol initiating system. The molar mass determined by SEC is $M_n = 800$, $M_w/M_n = 1.78$. In

6.2 Determination of Endgroups

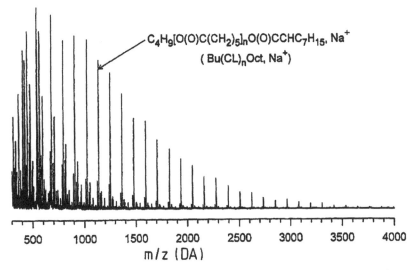

Fig. 6.21. MALDI-TOF spectrum of a poly-ε-caprolactone prepared with tin octoate in the presence of butanol, matrix: DHB. (Reprinted from [13] with permission of the American Chemical Society, U.S.A.)

the spectrum a series of signals dominate which can be assigned to the Bu(CL)$_n$Oct oligomers cationized with Na$^+$ ions.

A more detailed picture, presented in the fragment of the spectrum for the molar mass range 1000–1550 Da, reveals the formation of at least six different populations of macromolecules.

The assignments in the mass spectra correspond to the notation given below. All signals are due to [M+Na]$^+$ macromolecular ions, except for one series (B, K$^+$), which is indicated separately:

(A) Bu(CL)$_n$OSnOct   C$_4$H$_9$—[—OCO(CH$_2$)$_5$—]$_n$—OSnOCOCH(C$_4$H$_9$)(C$_2$H$_5$)

(B) Bu(CL)$_n$Oct   C$_4$H$_9$—[—OCO(CH$_2$)$_5$—]$_n$—OCOCH(C$_4$H$_9$)(C$_2$H$_5$)

(C) Bu(CL)$_n$OH   C$_4$H$_9$—[—OCO(CH$_2$)$_5$—]$_n$—OH

(D) H(CL)$_n$Oct   H—[—OCO(CH$_2$)$_5$—]$_n$—OCOCH(C$_4$H$_9$)(C$_2$H$_5$)

(E) H(CL)$_n$OH

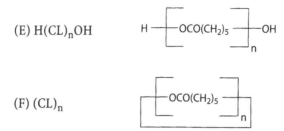

(F) (CL)$_n$

In Fig. 6.23 one of the signals from population A is expanded and shown separately by a bold line. It shows the typical isotope pattern for a Sn(II)-containing oligomer. As can be seen, it is over-

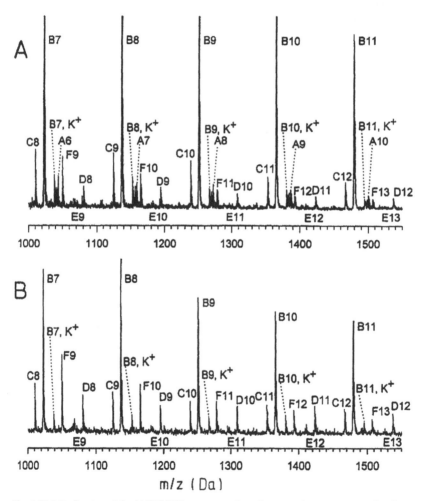

Fig. 6.22 A,B. Section of the MALDI-TOF spectrum of a poly-ε-caprolactone prepared with tin octoate in the presence of butanol: **A** under living conditions; **B** after treatment with hydrochloric acid, matrix: DHB. (Reprinted from [13] with permission of the American Chemical Society, U.S.A.)

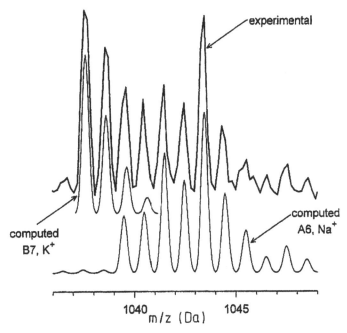

Fig. 6.23. Comparison of the 1036–1049 Da fragment of the MALDI-TOF spectrum of poly-ε-caprolactone from Fig. 6.21 with the computed isotopic distribution, matrix: DHB. (Reprinted from [13] with permission of the American Chemical Society, U.S.A.)

lapped partially with the signal of $B_7 + K^+$. A comparison with the computed isotopic distribution reproduces the experimental signal almost perfectly.

After hydrolysis of the reaction product with hydrochloric acid; see Fig. 6.22B, peak series A assigned to the Sn(II)-containing macromolecules disappeared. For the other populations the spectrum remains almost unchanged. This is additional evidence for the correct assignment of population A.

The commercially available tin octoate contains an admixture of water which may play the role of a co-initiator. Under conditions of low $Sn(Oct)_2/H_2O$ concentration only two populations of the PCL macromolecules are formed, namely E and F. In this case, Sn(II)-containing species cannot be observed; see Fig. 6.24A. However, when the starting concentration of $Sn(Oct)_2/H_2O$ is higher, an additional so far unobserved population F' can be identified; see Fig. 6.24B.

This population exhibits an isotopic distribution similar to population A and can be assigned to Sn(II)-containing cyclic oligomers:

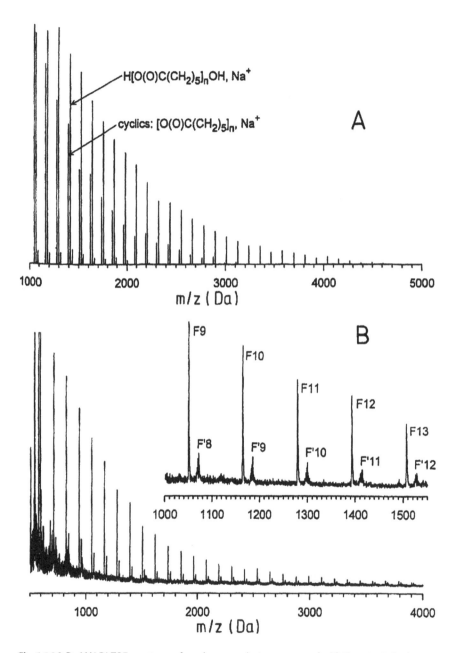

Fig. 6.24 A,B. MALDI-TOF spectrum of a poly-ε-caprolactone prepared with tin octoate in the presence of water: **A** Sn(Oct)$_2$ = 0.05 mol/l; **B** Sn(Oct)$_2$ = 1.0 mol/l, matrix: DHB. (Reprinted from [13] with permission of the American Chemical Society, U.S.A.)

6.2 Determination of Endgroups

Table 6.4. Applications of MALDI-TOF MS for endgroup analysis

Polymer	Matrix	Remarks	Reference
Epoxy resins	DHB	Analysis of endgroups in Bisphenol-A epoxy resins and epoxy-modified phenolic resols	[19]
Polymethyl methacrylate	DHB	Investigation of the polymerization of MMA with tert-butyllithium and triisobutylaluminium;	[20]
		Investigation of termination mode for thermal homopolymerization;	[25]
		Determination of chain transfer coefficients	[28]
Oligostyrene macromonomers	Dithranol	Functionality type analysis of methacryloyl endfunctionalized oligostyrene	[21]
Polyethylene glycol	DHB	Oligomer distribution, endgroups	[22,35]
	HABA	Oligomer distribution, endgroups	[23]
	DHB	Analysis of PEG with methacryloyl and formyl endgroups	[26]
Polyethylene oxide	α-cyano HCA	Oligomer distributions and endgroups of commercial nonylphenol ethoxylates	[32]
Aliphatic polyesters	IAA	Oligomer distribution, endgroups	[24]
Cyclic polyesters	DHB	Preparation and characterization of cyclic oligomers in aromatic esters	[27]
Polysulfide oligomers	9-nitro anthracene	Oligomer distribution, endgroups	[29]
Polystyrene	Dithranol, DHB	Analysis of TEMPO-capped polystyrene	[30]
		Analysis of fullerene end-capped polystyrene	[38]
Polythiophenes	Dithranol, 9-nitro anthracene	Analysis of fractions after solvent extraction	[31]
Polycarbonate	HABA, IAA	Endgroup analysis of unfractionated and fractionated Bisphenol-A polycarbonate	[33,34]
	DHB	Analysis of oligocarbonate diols	[39]
Polylactide	DHB	Investigation of the polymerization mechanism	[36]
Polyisobutylene	Dithranol	Functionality analysis of dihydroxy telechelics	[37]
Vinyl polyperoxides	DHB	Endgroup analysis	[40]
Polyoxymethylene	Dithranol	Endgroup analysis of trimethylsilylated and acetylated oligomers	[41]
Polethylene terephthalate	DHB	Analysis of PET degradation by plasma oxidation and hydrolysis	[42–44]

(F') $(CL)_n OSn$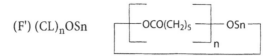

In conclusion, MALDI-TOF experiments reveal that in the polymerization of ε-caprolactone using tin octoate as the initiator macromolecules are formed having Sn(II) atoms in the polymer chain. The presence of the Sn atoms is a strong argument for a polymerization mechanism proceeding with Sn-alkoxides as active species. This is a nice example for the use of MALDI-TOF in the elucidation of reaction mechanisms.

6.2.6 Further Applications

In recent years MALDI-TOF became the premier technique for determining functional groups, in particular endgroups, in complex macromolecular systems. Its high sensitivity and selectivity makes it possible to investigate polymerization mechanisms by analyzing the formation and reaction of functional groups as well as to determine defective structures that are formed as a result of secondary reactions. A number of interesting applications has been reviewed in [1]. Some further applications in this field are summarized in Table 6.4.

6.3 Analysis of Copolymers

When two or more monomers of different chemical structures are involved in a polymerization reaction, instead of a chemically homogeneous homopolymer, in most cases a chemically heterogeneous copolymer is formed. Depending on the reactivity of the monomers and their sequence of incorporation into the polymer chain, macromolecules can be formed which differ significantly in composition. (meaning the amounts of repeat units A, B, etc. in the copolymer) and the sequence distribution. With respect to sequence distribution copolymers can be classified as alternating, random, block, and graft copolymers.

Consider a random copolymer which has been obtained in a homogeneous reaction from a mixture of A and B monomers. Even under such favourable conditions the resulting macromolecules will differ in chemical structure. There are differences in the sequence of the A and B monomers along the macromolecules, differences in the average chemical composition of the copolymer molecules formed at any instant of the polymerization, and differences due to the depletion of the reaction mixture in one of the monomers.

The average sequence lengths for the A- and B-units can be measured by physical or chemical methods. The former (FTIR or NMR analyses) usually measure the percentage of A and B units inside of triads, pentades etc. whereas the latter methods evaluate the percentage of A-A, A-B, and B-B linkages. Macromolecules of random copolymers, even if identical in chain length and composition (and thus also in the average sequence length), still offer a great variety with respect to the order of individual sequences in the molecules. Thus, a copolymer sample contains a tremendous number of constituents differing in chain length and composition.

MALDI-TOF offers the opportunity to determine the composition of single copolymer molecules precisely through accurate mass measurements. With the mass resolution of state-of-the-art MALDI-TOF instrumentation it is possible to resolve mass peaks of single oligomers up to masses of about 40 to 50 kD. Such measurements should allow structural characterization of copolymers including the monomer ratio and the identification of terminal groups.

6.3.1 Diblock Copolymers of α-Methylstyrene and 4-Vinylpyridine [45]

Block copolymers have many interesting properties with applications as compatibilizers and phase stabilizers for polymer blends, applications to drug delivery systems, micellation and brush formation. Very frequently, the end-use properties of block copolymers not only depend on the total molar mass and the chemical composition but also on the size ratio of the individual blocks.

Aim

Using the classical techniques like FTIR and NMR spectroscopy or SEC it is not possible to determine the molar mass distributions of both blocks in a block copolymer. In general, block copolymers are prepared by sequential polymerization forming a block of poly-A first, followed by the formation of the second block poly-B to give the block copolymer polyA-*b*-polyB. Since a fraction can be taken out of the reaction mixture after the formation of poly-A, this block can be analyzed with regard to molar mass distribution by SEC. This, of course, cannot be done for the second block and, therefore, its average molar mass is usually calculated by substracting the molar mass of poly-A from the total molar mass of the block copolymer.

In the present study, MALDI-TOF shall be used to determine the molar mass distributions of both blocks in the block copolymer. The subject of study are diblock copolymers of α-methylstyrene and 4-vinylpyridine which are of considerable interest in micelle formation. Each of the block copolymers has an identical length of the first segment and various lengths of the second block. This allows for the study of the sequential anionic polymerization process.

Materials

The poly(α-methylstyrene)-*b*-poly(4-vinylpyridine) diblock copolymers (PαMS-*b*-P4VP) were prepared by sequential anionic polymerization in THF at −78 °C. As the initiator *sec*-butyl lithium was used. The homopolymer PαMS was recovered by extracting a sample before the addition of the second monomer. To synthesize the block copolymers, 4VP monomer was added in three steps in order to obtain block copolymers with the same length of PαMS but different lengths of P4VP. The block copolymers were isolated by precipitation in hexane and dried under vacuum. The characteristics of the block copolymer samples are given in Table 6.5.

Equipment

MALDI-TOF system — PerSeptive Biosystems Voyager-DE and Voyager-DE STR instruments, acceleration voltage of positive 20 kV with delayed extraction, detection in linear mode, 250 transients were summed per full spectrum

MALDI-TOF sample preparation — Samples were prepared with dithranol (10 mg/ml) as the matrix, silver trifluoroacetate (1 mg/ml) as the cation donor, and dimethyl formamide as the solvent. The polymer samples (1 mg/ml) were mixed 20:20:1 v/v/v with matrix/polymer/salt solution.

MALDI-TOF Analysis

The MALDI-TOF spectra of the block copolymers consist of numerous peaks, whose number and complexity grow with the molar mass of the sample; see Fig. 6.25.

As a result of the initiation with *sec*-butyl lithium, every molecule of the block copolymers consists of a butyl and a hydrogen

Table 6.5. Characteristics of PαMS-b-P4VP diblock copolymers

Sample	Monomer volumes ratio (αMS/4VP)	PαMS-*b*-P4VP[a]	PαMS-*b*-P4VP[b]
PVP1	10/0	11.1-*b*-0	11.1-*b*-0
PVP2	10/1	11.1-*b*-1.4 (+/− 0.4)	----
PVP3	10/4	11.1-*b*-6.1 (+/− 1.3)	----
PVP4	10/8	11.1-*b*-13.0 (+/− 4.4)	----

[a] Block lengths are calculated from the monomer volume ratios, assuming 11.1 units for the PαMS block
[b] Determined by SEC, the block copolymers could not be analyzed by SEC

6.3 Analysis of Copolymers

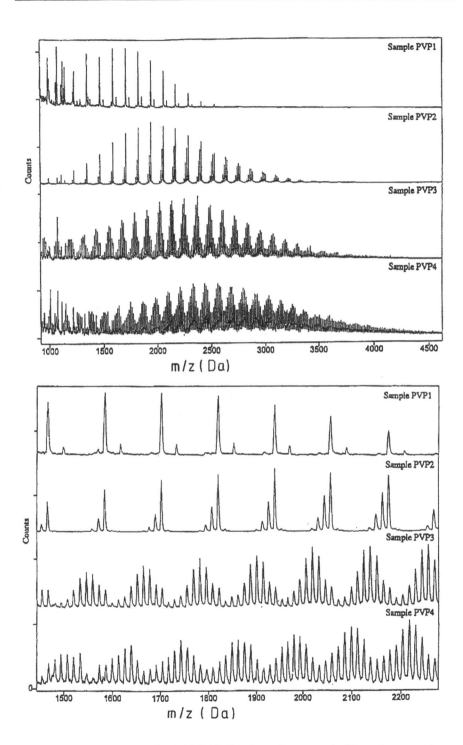

Fig. 6.25. MALDI-TOF spectra of PαMS-*b*-P4VP diblock copolymers, matrix: dithranol; *top*: complete spectra, *bottom*: enlarged details (Reprinted from [45] with permission of Wiley, U.S.A)

endgroup, and numbers x and y of αMS and 4VP units, respectively. The ionization of the macromolecules takes place through the attachment of one silver cation per molecule to form $[M+Ag]^+$ molecular ions. The m/z values can then be calculated according to the following equation:

$$m/z = [M+Ag]^+ = 57.115(Bu) + 1.008(H) + 118.18x + 105.14y \quad (6.4)$$

where 118.18 and 105.14 are the masses of the αMS and 4VP units, respectively.

For copolymers, the key problem is to find a combination of x and y values that match the experimental value of m/z for the most intense peak of the spectrum. As there are two variables, there are many solutions possible. Therefore, additional information on the chemical composition of the sample is required. This information can come from a knowledge of the monomer feed or from the determination of the average composition of the copolymer by spectroscopic means. Possible assignments of copolymer composition for the most intense peaks in each spectrum are given in Table 6.6.

As can be seen in the table, the differences between the experimental and the calculated masses vary depending on the monomer ratio. One could now carry out an assignment based on the minimum value for Δ. On the other hand it is known that mass differences of 2 Da are acceptable under certain conditions. For example, the calibration procedure can affect the position of the mass peak. Furthermore, the width of the mass peaks is rather high, reaching 5 Da in some cases. This is due to the underlying isotope distribution which is additionally broadened by 2 units due to the silver cation.

Another, more rational assignment is based on the values that are closest to the total chemical composition of the samples.

Table 6.6. Possible assignments for mass peaks of the block copolymers

Sample	$[M+Ag]^+$ exp.	$[M+Ag]^+$ calc.	x	y	Δ = exp-calc
PVP1	1702.31	**1702.33**	13	0	−0.02
PVP2	1938.69	1939.51	7	9	−0.82
		1938.69	**15**	**0**	**0.00**
PVP3	2492.49	**2490.47**	17	3	2.02
		2491.29	9	12	1.20
		2492.11	1	21	0.38
PVP4	2558.60	**2556.49**	14	7	**2.11**
		2557.31	6	16	1.29

These are marked in bold in Table 6.6. Since peak analysis is carried out for the most intense peaks in the spectra these are expected to be close to the average composition.

After this rationalizing step [46], the chemical composition of the other peaks can be assigned. For this assignment two properties of the mass spectrum are used:

1. When shifting from the most intense peak of one peak cluster to the most intense peak of the next cluster, the number of 4VP segments (y) remains constant, while the number of segments of αMS (x) increases by one.
2. Within a cluster of peaks, the number of segments of the heavier monomer αMS (x) increases with increasing mass, while the number of segments of the lighter monomer 4VP (y) decreases. The total number of segments x+y is constant within each cluster.

Further refinement of the assignment procedure is introduced by using a statistical random coupling hypothesis test. Using this approach, the chance of false assignments can be reduced to a minimum.

The assignment of the peaks in the mass spectra of samples PVP1 to PVP4 is given in Fig. 6.26. For sample PVP1 this assignment is straightforward because this sample represents the PαMS homopolymer. Also, sample PVP2 can be analyzed easily. In the lower molar mass range, clusters of three to four mass peaks are obtained, which can be assigned to PαMS homopolymer, and copolymer molecules with one to three 4VP segments in the polymer chain.

As the number of peaks within a cluster increases with increasing numbers of 4VP segments in the macromolecules, assignment of the mass peaks for samples PVP3 and PVP4 is more complicated. The last peaks of one cluster tend to overlap with the first peaks of the next cluster and therefore assignment can become inaccurate due to multiple assignment possibilities, see PVP4 in Fig. 6.26. Again, a mathematical treatment of the data is required (which shall not be discussed here) to find the most probable assignment.

After the assignment of all mass peaks, the relative intensities of the mass peaks can be analyzed to obtain semiquantitative information on the complex structure of the samples. In the present case it is assumed that the ionizability of the macromolecules is not a function of the chemical composition, but all molecules are ionized with the same probability. The normalized intensities of the peaks form an array $N_{exp}(x,y)$ which can be presented in a

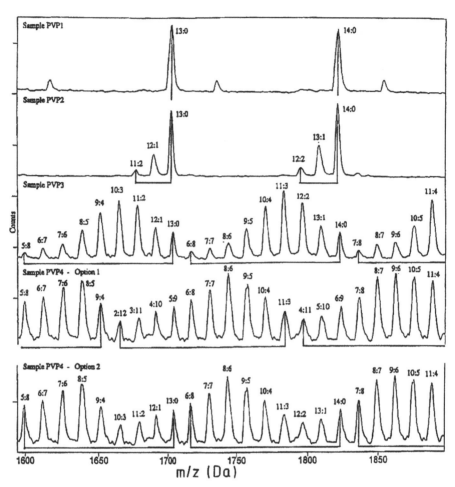

Fig. 6.26. Example of the assignment of chemical composition, numbers indicate the ratio x:y, peaks belonging to the same cluster are *underlined*. (Reprinted from [45] with permission of Wiley, U.S.A.)

three-dimensional surface plot of the experimental length distributions of the two blocks in the copolymer; see Fig. 6.27.

These three-dimensional plots give a view on all stages of the polymerization under study. Figure 6.27A,B clearly shows how the homopolymer PαMS gets transformed into the diblock copolymer PαMS-*b*-P4VP. It is obvious that the majority of sample PVP2 is still composed of the homopolymer. On the other hand, the distribution of the 4VP units can be nicely obtained from the plot. Upon further addition of 4VP monomer the homopolymer is gradually consumed and disappears completely in sample PVP4. At the same time, the lengths of the 4VP blocks and their distributions increase. Therefore this technique provides a unique insight

6.3 Analysis of Copolymers

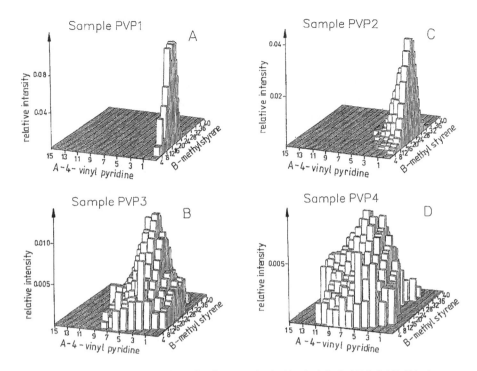

Fig. 6.27 A–D. Experimental normalized distributions of units, $N_{exp}(x,y)$, for PαMS-b-P4VP diblock copolymer, samples: **A** PVP1; **B** PVP3; **C** PVP2; **D** PVP4. (Reprinted from [45] with permission of Wiley, U.S.A.)

into the polymerization process, which is impossible to obtain by any other means.

From the single molecule concentrations a number of structural properties, including the average chain lengths of the blocks, their polydispersity, and the amount of PαMS homopolymer can be calculated. These numbers are presented in Table 6.7 for the different copolymers.

As a proof for the validity of the present approach a comparison was made between the data obtained by the MALDI-TOF meas-

Table 6.7. Structural properties of the block copolymers calculated from MALDI-TOF data (polydispersities given in parentheses)

Sample	Amount PαMS (mol%)	x in PαMS	x in PαMS-b-P4VP	y in PαMS-b-P4VP
PVP1	100	12.3 (1.08)	---	---
PVP2	42	16.2 (1.07)	17.8 (1.05)	1.8 (1.29)
PVP3	5	15.0 (1.08)	15.8 (1.16)	4.2 (1.27)
PVP4	0	---	14.8 (1.20)	7.6 (1.12)

urements and other independent methods. Table 6.8 presents the comparison with data from NMR and the feed composition of the reaction mixture. As can be seen, the PαMS content is shifted towards higher than expected values increasing with the molar mass of the copolymers. Possible reasons for this discrimination at higher masses can be lower desorption efficiency, ionization cross section, or detection efficiency.

Table 6.8. Comparison of the chemical composition of the copolymers obtained by different experimental techniques (mole fraction of PαMS)

Sample	Molar ratio of reagents	NMR	MALDI-TOF
PVP2	0.9±0.1	0.91	0.94
PVP3	0.63±0.06	0.66	0.81
PVP4	0.45±0.04	0.41	0.69

6.3.2 Copolymers of Ethylene Oxide and Propylene Oxide [47]

Aim

Copolymers of ethylene oxide (EO) and propylene oxide (PO) are of significant industrial importance. They are used as surfactants and as precursors for the production of polyurethanes. The MALDI-TOF analysis of EO and PO homopolymers has been discussed by a number of authors. Even using standard procedures it is not a major problem to determine the type of endgroups and the degree of polymerization; see Sect. 7.3.3.

The analysis of EO-PO copolymers is much more complicated mainly due to the fact that multiple different combinations of EO and PO units in the polymer chain result in very complex spectra. The low mass difference between the EO unit (44.03 Da) and the PO unit (58.06 Da) gives rise to multiple peak overlappings even at low molar masses. Accordingly, for a proper analysis of these copolymers, high resolution of the MALDI-TOF instrument is of paramount importance.

The aim of the present study is to produce well resolved spectra of EO-PO copolymers which can be analyzed with regard to EO-PO composition. Instrumentation with a mass accuracy better than 0.01% is used. Resolution of approximately 2600 is achieved for isotope separation across the entire distribution. In a second part, the effect of experimental conditions on the quantitative compositional analysis is discussed.

Materials

The EO-PO copolymers under study were technical products of Aldrich Chemical Co. or Nalco Chemical Co. The Aldrich copolymer was a [PPO-b-PEO-b-PPO]-bis(2-aminopropyl ether).

Equipment

MALDI-TOF system — custom-made time-lag focusing instrument described in [48] with a 337-nm nitrogen laser, acceleration voltage of positive 30 kV, 50–100 transients were summed per full spectrum.

MALDI-TOF sample preparation — Samples were dissolved in 1,4-dioxane at a concentration of 5 mg/mL, HABA was used as the matrix as 0.05 molar solution in 1,4-dioxane. Polymer stock solutions were diluted tenfold with the matrix solution and a small amount of a 0.01 molar NaCl aqueous solution was added. Then 1 μl of the mixture was added to the MALDI-TOF probe tip and allowed to air-dry.

MALDI-TOF analysis

The first sample under investigation is a EO-PO triblock copolymer of the following general composition:

$$NH_2CH(CH_3)CH_2[PO]_x[EO]_y[PO]_zNH_2$$

The MALDI-TOF spectrum that is obtained for this polymer is given in Fig. 6.28. In the present case, unit mass resolution across the displayed mass range is achieved. As can be seen from the spectrum, there is a pattern of three isotop clusters that repeats every 44.03 Da, from which the presence of ethylene glycol units can be postulated.

However, the identities of the other monomer and the endgroups are not directly obvious from the spectrum. To elicit compositional information, one requires information on the masses of the endgroups and the copolymerized monomers. Accordingly, high mass accuracy and resolution are of great importance. In the present instrument, mass accuracy of better than 0.01 % is achieved.

With this high accuracy, a single peak representing a monoisotopic mass can be selected from one of the isotope clusters. By calculating oligomer masses for different oligomer compositions, a possible peak assignment can be carried out. For example, the mass peak at 1842.12 Da has two possible compositions that fall within the accuracy of the mass determination:

1. 37 EO units + 2 PO units: calculated $[M+Na]^+ = 1842.13$ Da
2. 8 EO units + 24 PO units: calculated $[M+Na]^+ = 1842.29$ Da

Here it cannot be decided directly, which assignment is correct. However, one can consider a more extensive mass range, as is shown in Table 6.9.

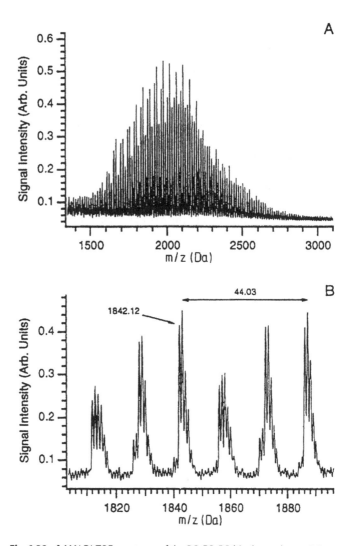

Fig. 6.28. **A** MALDI-TOF spectrum of the PO-EO-PO block copolymer. **B** Expanded region. (Reprinted from [47] with permission of the American Chemical Society, U.S.A.)

This table shows that for a series of peaks with a mass difference of one EO unit, one composition is favored over the other. The spectrum can be checked for self-consistency as a test for the validity of a certain assignment. The (37EO+2PO) composition is confirmed through inspection of the rest of the distribution, as each peak in the spectrum can be assigned using this composition as a starting point. This is not the case for the (8EO+24PO) composition.

Table 6.10 contains the peak assignments for a typical repeat pattern that is found in the spectrum of Fig. 6.28B. These assign-

Table 6.9. Comparison of experimental and calculated mass data for several peaks in Fig. 6.28

Exp. mass	EO/PO	Calc. mass	Accuracy (%)	EO/PO	Calc. mass	Accuracy (%)
1798.03	36/2	1798.10	0.004	7/24	1798.26	0.013
1842.12	37/2	1842.13	0.001	8/24	1842.29	0.005
1886.12	38/2	1886.15	0.002	9/24	1886.31	0.010
1930.18	39/2	1930.18	0.000	10/24	1930.34	0.008
1974.16	40/2	1974.21	0.002	11/24	1974.37	0.010
2018.11	41/2	2018.23	0.006	12/24	2018.39	0.014
2062.11	42/2	2062.26	0.008	13/24	2062.42	0.015
2106.10	43/2	2106.29	0.009	14/24	2106.45	0.016

Table 6.10. Assignment of peaks detected for a typical repeat unit shown in Fig. 6.28B

Measured mass	EO	PO
1856.14	36	3
1858.11	40	0
1870.13	35	4
1872.15	39	1
1886.15	38	2

ments identify the monoisotopic form of the indicated oligomers and provide an appreciation for the problem of copolymer analysis by MALDI-TOF. Within a 30 Da mass window, five oligomers are represented. This spectral congestion would only worsen with increased compositional heterogeneity. Note that sequence information is beyond the capability of MALDI-TOF. Therefore, while it is possible to conclude that the sample is a copolymer and not a blend of two homopolymers, it cannot be determined if the copolymer is random or block.

Due to the high resolution of the MALDI-TOF instrument, isotope separation across the entire distribution is achieved and the mass accuracy is sufficient for unique peak identifications. Still, a computer-assisted approach would be useful, where spectral simulations are performed and fit to the experimental spectrum to obtain the relative intensities of the assigned peaks. Various compositional iterations could be assessed to arrive at the best fit for the data. Such an approach is essential for the extraction of relative oligomer compositions and overall average polymer composition.

The situation becomes more complex with an increase in the molar mass of the polymer and the compositional heterogeneity. This is shown in Fig. 6.29 for a technical EO/PO copolymer. The expansion of the spectrum for different mass regions shows that only in the low-mass region is sufficient resolution obtained to perform a single isotope analysis. Where there is still unit mass resolution in the high-mass region, the opportunity to determine the oligomer composition exists but the isotopic envelopes begin to include two or more possible oligomer compositions.

In addition to the resolution of the MALDI-TOF instrumentation other experimental factors are of great importance for the compositional analysis of EO/PO copolymers. The effects of sample concentration, laser power, type of matrices and solvents have been illustrated by Chen et al. [49]. They showed that great care needs to be exercised when interpreting copolymer spectra for compositional analysis, even for copolymers with structurally similar monomers. When using different matrices, the deviations in relative peak ratios can be larger than 30%.

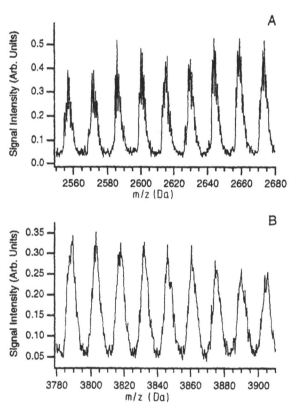

Fig. 6.29 A,B. Expansions of the MALDI-TOF spectrum of a technical EO/PO copolymer: **A** low-mass region; **B** high-mass region. (Reprinted from [47] with permission of the American Chemical Society, U.S.A.)

6.3.3 Modified Silicone Copolymers [50]

Modified polydimethyl siloxane (PDMS) is used in the cosmetics industry to modify the properties of cosmetic formulations. Different functional groups such as ethoxy, perfluoroalkyl, and others are used to impart a feeling of delicacy. Although the manufacturing methods are well characterized, and the structures of the products can be estimated from the process variables and analytical results, detailed structural characterization is required in order to correlate the structures with subjective responses to cosmetics and to improve performance. In the present application, the analysis of perfluoroalkyl modified PDMS shall be discussed. *Aim*

The modified silicone copolymers were technical products of Shinetsu Chemical Co. Ltd, Tokyo, Japan. Their compositions and molar masses are summarized in Table 6.11. *Materials*

$$H_3C-\underset{\underset{CH_3}{|}}{\overset{\overset{CH_3}{|}}{Si}}-O-\left[\underset{\underset{CH_3}{|}}{\overset{\overset{CH_3}{|}}{Si}}-O\right]_m-\left[\underset{\underset{R}{|}}{\overset{\overset{CH_3}{|}}{Si}}-O\right]_n-\underset{\underset{CH_3}{|}}{\overset{\overset{CH_3}{|}}{Si}}-CH_3$$

MALDI-TOF system — Kompact MALDI III (Kratos-Shimadzu) with a 337-nm nitrogen laser, acceleration voltage of positive 20 kV, reflectron mode, each mass spectrum was the average of 100 shots. *Equipment*

MALDI-TOF sample preparation — Samples were dissolved in chloroform at a concentration of 0.5% (wt/vol), DHB was used as the matrix as 100 mg/ml solution in acetone. Sodium perchlorate was used as cation donor and dissolved in acetone at a concentration of 0.1 µmol/µl. Aliquots of 50 µl of the sample solution, 50 µl of the matrix solution, and 1 µl of the salt solution were mixed. A

Table 6.11. Properties of the modified silicone copolymers

Sample	m/n[a]	R	M_w[b] (g/mol)
F1	9/1	$CH_2CH_2CF_3$	3900
F3	6/4	$CH_2CH_2CF_3$	4020
F4	9/1	$CH_2CH_2C_4F_9$	4110
F5	8/2	$CH_2CH_2C_4F_9$	4600
F6	7/3	$CH_2CH_2C_4F_9$	4400

[a] Based upon the amounts formulated
[b] Determined by size exclusion chromatography

MALDI-TOF analysis

0.7 μl aliquot of the mixed solution containing about 2 μg of the sample was pipetted onto the sample slide and air-dried.

The MALDI-TOF spectra of samples F1 and F3 are shown in Fig. 6.30. As indicated in Table 6.12, the molar masses obtained by SEC are in the range from 4000 to 6000 g/mol. The mass peaks in the MALDI-TOF spectra, however, are centered around 1300 Da. This is obviously due to the fact that the polydispersity of the samples is larger than 1.1 to 1.2. For a proper determination of average molar masses of the samples by MALDI-TOF a SEC prefractionation must be carried out as has been proposed by Montaudo et al. [51].

The analysis of the mass peaks is conducted by comparing the measured with the calculated peak masses for oligomers of different chemical composition. It is assumed that all oligomers are ionized through the attachment of a sodium cation forming the corresponding $[M+Na]^+$ macromolecular ions. The peak masses are then calculated using the following equation:

$$[M+Na]^+ = 22.990(Na) + 162.381(endgroups) + 74.155m + 156.179n \quad (6.5)$$

where m and n are the numbers of dimethyl siloxane (DMS) and methylperfluoroalkyl siloxane units (MFAS), accordingly.

Since the mass differences between the repeat units DMS and MFAS are sufficiently large, the assignment of the mass peaks in Fig. 6.30 is rather straightforward. For these samples dual or multiple assignments are not found. In all cases, the major peak series exhibit peak-to-peak mass increments of 74 Da with correspondance to the DMS repeat unit. This is in agreement with the expectations since the m/n ratio is significantly larger than one for all samples; see Table 6.11.

The peak assignments for samples F1 and F3 are indicated as (m,n) in Fig. 6.30. As can be seen, the major peak series in sample F1 is that of PDMS (n=0). A second series of lower intensity can be assigned to PDMS where three units of MFAS are incorporated (n=3). m is found to be 8–33. In sample F3, the peak series for PDMS disappears, while the series with n=3 becomes the main component. For this sample a second peak series is detected that can be assigned to a PDMS with six incorporated units of MFAS (n=6). The range of m is 5–17 in this case. No peaks for PDMS with n=1,2,4, or 5 are detected.

The unusual fact that only values of n=0,3,6 can be detected in the spectra indicates that it is not the monomeric MFAS that has been used for the preparation of the samples. Instead, it can be assumed that an MFAS trimer was used similar to the DMS trimer that is frequently used in silicone synthesis.

6.3 Analysis of Copolymers

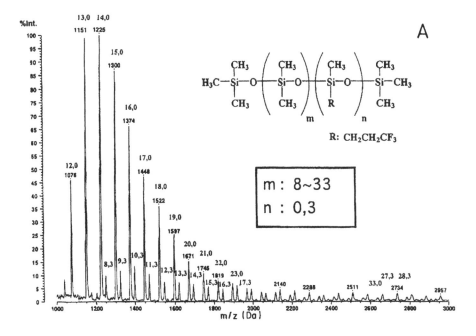

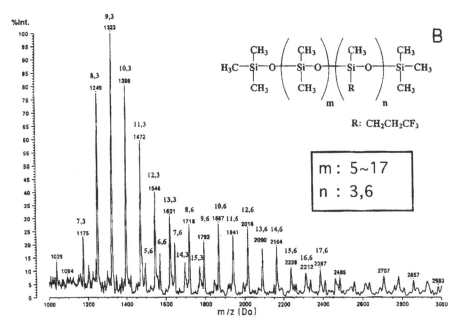

Fig. 6.30 A,B. MALDI-TOF spectra of perfluoroalkyl modified silicone copolymers: **A** F1; **B** F3. (Reprinted from [50] with permission of Wiley, UK)

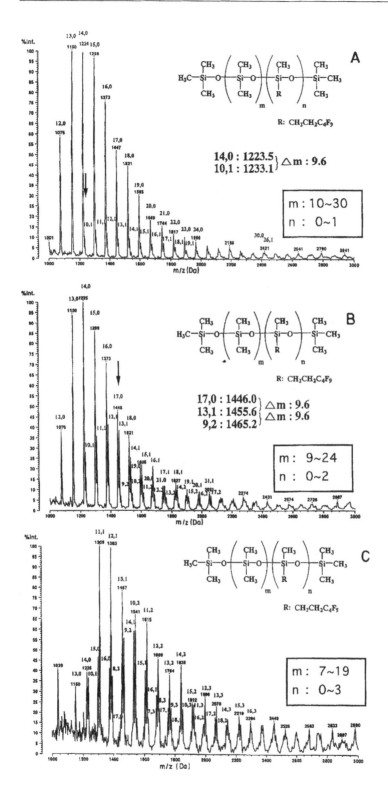

6.3 Analysis of Copolymers

A different behavior is obtained for samples F4-F6 where instead of CF_3 substituents C_4F_9 groups are present. The MALDI-TOF spectra of these samples are presented in Fig. 6.31. The calculation of the peak masses for the different oligomers is done using the following equation:

$$[M+Na]^+ = 22.990\,(Na) + 162.381\,(endgroups) + 74.155\,m + 306.200\,n \quad (6.6)$$

where n is the number of the corresponding MFAS unit.

Similar to the previous series of samples, in F4 the major peak series constitutes PDMS without MFAS units (n=0). Different from the previous series, however, a second peak series PDMS with one MFAS unit per molecule (n=1) is detected. In sample F5, peak series for n=0, 1 and 2 are obtained, while in sample F6 n is extended to 0-3. The mass difference between peaks of different oligomer series is 9.6 Da, but resolution is still sufficient for detecting all peak series. The range of m is 10-30 for sample F4, 9-24 for sample F5, and 7-19 for sample F6.

Comparing sample series F1-F3 with F4-F6 it is obvious that the latter is not prepared from the MFAS trimer, because n takes all values from 0 to 3.

6.3.4 Monitoring the Sulfonation of Polystyrene [52]

Aim Copolymers cannot only be produced by direct copolymerization. Very frequently, a copolymer is obtained when a homopolymer is modified by a polymer analogous reaction. One of the very typical examples for this approach is the production of ethylene alcohol-vinyl acetate copolymers by partial hydrolysis of polyvinyl acetate.

In the present application, polystyrene shall be functionalized via sulfonation. In a second step, the polystyrene sulfonic acid shall be transferred into the corresponding sodium salt. The degrees of both reactions will be investigated by MALDI-TOF mass spectrometry and compared to data obtained by SEC.

Materials The parent polystyrenes were synthesized by anionic polymerization with butyl lithium as the initiator. The corresponding polystyrene sulfonic acids were prepared from these polystyrenes in a mixture of concentrated sulfuric acid and phosphorous pentoxide. The free acid form was prepared by passing the sodium salt through a column of Amberlyst cation exchanger. The sodium

◄ **Fig. 6.31 A-C.** MALDI-TOF spectra of perfluoroalkyl modified silicone copolymers: **A** F4; **B** F5; **C** F6. (Reprinted from [50] with permission of Wiley, UK)

salts and the free acids were dialyzed in water prior to use. The samples were products of Polymer Standards Service GmbH, Mainz, Germany. The general reaction scheme was as follows:

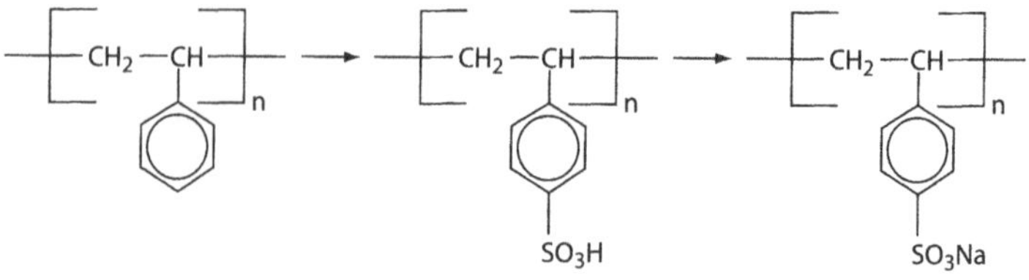

$$-[CH_2-CH(C_6H_5)]_n- \longrightarrow -[CH_2-CH(C_6H_4SO_3H)]_n- \longrightarrow -[CH_2-CH(C_6H_4SO_3Na)]_n-$$

      1a   1b   1c

Equipment

MALDI-TOF system — Bruker REFLEX equipped with a 337 nm nitrogen laser, acceleration voltage was 15–35 kV, spectra were taken in linear and reflectron modes, each mass spectrum was the average of 100–200 shots.

MALDI-TOF sample preparation — Samples were dissolved in THF. DHB, sinapinic acid, and 5-chlorosalicylic acid were used as the matrices. They were dissolved in water and THF, correspondingly, at a concentration of 0.01 mol/l. Silver trifluoroacetate was used as cation donor. Polymer solutions were mixed with the matrix at a mole ratio of 1:0.001 (matrix:polymer). 0.5–1 µl of the sample solution was pipetted onto the sample slide and air-dried.

MALDI-TOF analysis

In the present application samples are investigated that show single chain resolution and allow a detailed analysis of individual molecules in the polymer mixture. Their molar masses range from 2000 to 10,000 g/mol. The MALDI-TOF spectra of the series with the lowest molar mass are shown in Fig. 6.32.

Figure 6.32A shows the positive ion mass spectrum of the parent polystyrene 1a. The spectrum shows single chain resolution which allows one to determine the molar masses of individual polymer ions. The peak-to-peak distance is 104 Da which is the mass of the polystyrene repeating unit, and the absolute peak masses fit to molecules with a butyl end group and an attached silver cation. Only one oligomer series is obtained in the spectrum indicating that the present sample is uniform in chemical composition:

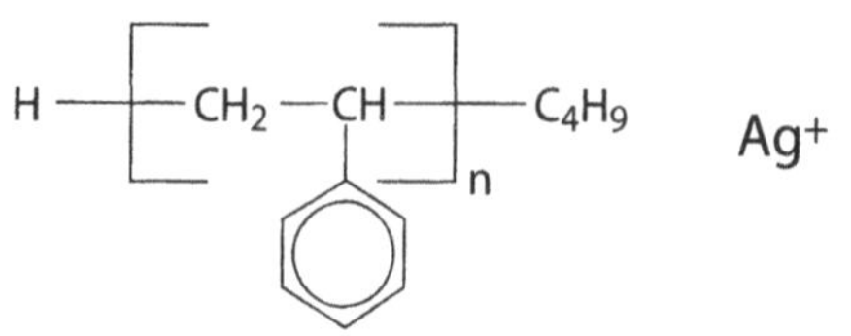

$$H-[CH_2-CH(C_6H_5)]_n-C_4H_9 \quad Ag^+$$

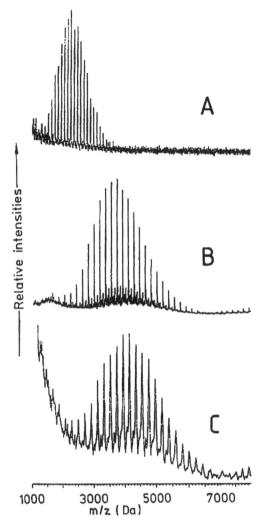

Fig. 6.32A–C. MALDI-TOF spectra of **A** the parent polystyrene **1a**: with $M_p=2140$; **B** polystyrene sulfonic acid **1b** with $M_p=3740$; **C** the corresponding sodium salt **1c** with $M_p=4185$. (Reprinted from [52] with permission of Wiley-VCH, Germany)

The mass spectrum of the polystyrene sulfonic acid **1b** which was prepared by soft sulfonation from sample **1a** is shown in Fig. 6.32B. In this case the peak-to-peak mass increment is 184 Da and corresponds to the polystyrene sulfonic acid repeat unit. The absolute peak masses coincide with fully sulfonated polystyrene molecules. This means that each aromatic subunit carries at least one sulfonic acid group. The polymer masses can be calculated according to the following formula; for sample **1b** ionization takes place via the abstraction of a proton and the formation of singly charged negative ions:

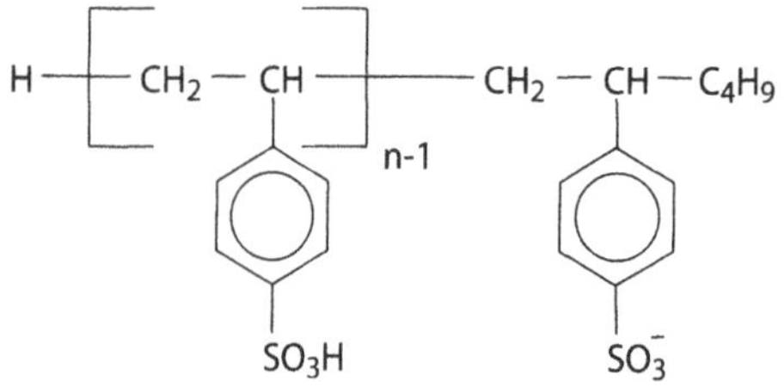

$$H-[-CH_2-CH-]_{n-1}-CH_2-CH-C_4H_9$$
(with phenyl-SO$_3$H on the repeat unit and phenyl-SO$_3^-$ on the end group)

Apart from the most intense peak series, additional small peaks with the same peak-to-peak mass increment of 184 Da are obtained; see Fig. 6.33. These additional peaks are due to different degrees of sulfonation. The mass difference towards the major peak series is −80 Da, which is equivalent to the loss of one sulfonate group. Accordingly, the mass peak at 3661 Da corresponds to an oligomer series with a deficit of one sulfonate group, while the series with a peak at 3581 Da indicates a deficit of two sulfonate groups.

The oligomer masses can be calculated based on the following structure:

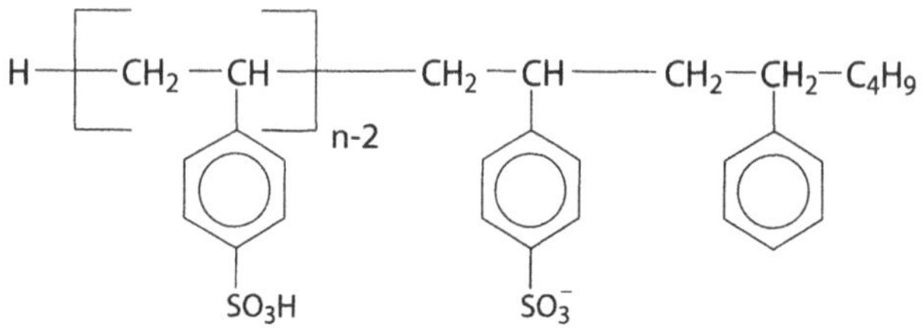

$$H-[-CH_2-CH-]_{n-2}-CH_2-CH-CH_2-CH_2-C_4H_9$$
(with phenyl-SO$_3$H, phenyl-SO$_3^-$, and phenyl end groups)

Sample 1c is the sodium salt of the polystyrene sulfonic acid 1b; see Fig. 6.32C. The peak series of this sample has a peak-to-peak mass increment of 206 Da and corresponds to the repeating unit of the salt. Again, negative molecular ions are detected which are formed through the loss of one sodium cation:

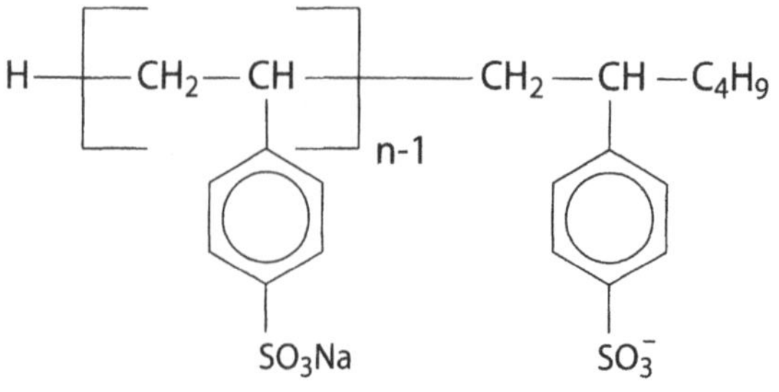

$$H-[-CH_2-CH-]_{n-1}-CH_2-CH-C_4H_9$$
(with phenyl-SO$_3$Na on the repeat unit and phenyl-SO$_3^-$ on the end group)

6.3 Analysis of Copolymers

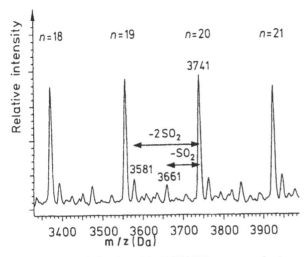

Fig. 6.33. Expanded region of the MALDI-TOF spectrum of polystyrene sulfonic acid **1b** with incomplete sulfonation. (Reprinted from [52] with permission of Wiley-VCH, Germany)

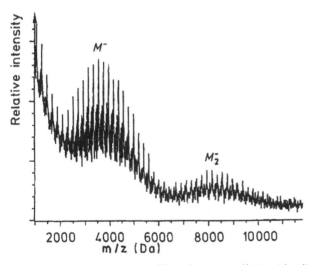

Fig. 6.34. MALDI-TOF spectrum of the polystyrene sulfonic acid sodium salt **1c** with distribution of dimeric clusters. (Reprinted from [52] with permission of Wiley-VCH, Germany)

An interesting feature of this spectrum is the formation of singly charged dimeric clusters, which can be assigned to the sum of an intact polymer chain with all sodium counterions plus a polymer chain with a single missing sodium counterion; see Fig. 6.34.

Table 6.12. Molar mass analysis of the samples 1a–c (g/mol)

	1a	1b	1c
M_p (MS)	2140	3740	4160
n	20	20	20
M_p (SEC)	2380	------	3800
M_n (MS)	2280	2870	4270
M_w (MS)	2380	4090	4520

After baseline correction, all spectra are characterized by the mass at peak maximum (M_p), the number-average molar mass (M_n), and the weight-average molar mass (M_w). In the case of the parent polystyrenes and the sodium salts of the polystyrene sulfonic acid, the SEC values of the manufacturer are also included; see Table 6.12.

Based on the peak molar mass of the parent polystyrene and the knowledge of the ionized species one can calculate the expected peak maxima of the corresponding acid and sodium salt assuming 100% conversion. For sample 1b the calculated M_p is 3741 Da which is very close to the experimentally determined value. Also, the calculated peak molar mass for the sodium salt, which is 4159 Da, fits the experimental data perfectly. This indicates that both reaction steps occurred with nearly 100% conversion.

As can be seen in Fig. 6.33, the peaks of the incompletely sulfonated oligomers are very small. Comparing their peak intensities with the peak intensities of the fully sulfonated oligomers, a ratio of the peak areas of 2:98 is obtained, which fully supports the assumption of nearly 100% conversion in the sulfonation reaction.

6.3.5 Modified Phenol-Formaldehyde Resins [53]

Aim

Condensation polymers, such as phenolic resins, are important technical products. Although they are rather low in molar mass, compared to polymerization products, they exhibit a complex polymer structure in many cases. In addition to the molar mass distribution, they may be distributed in chemical composition, due to the formation of different monomer sequences along the polymer chain. From studies on polycondensation kinetics and the modelling of the individual steps of the synthesis it is known that resin molecules with different endgroups may be formed. Accordingly, when describing the molecular heterogeneity of condensation polymers, their chemical composition and endgroup

6.3 Analysis of Copolymers

functionality must be considered in addition to molar mass distribution.

For fine-tuning of application properties, phenolic resins are modified frequently with other monomers like urea and melamine. The resulting products are then copolymers of phenol-urea-formaldehyde (PUF) and phenol-melamine-formaldehyde (PMF), accordingly. The analysis of such complex resins by MALDI-TOF will be dealt with in the following application.

The samples under investigation are commercial products of Bakelite AG, Duisburg, Germany.

Materials

MALDI-TOF system — Kratos Kompact MALDI III equipped with a 337-nm nitrogen laser, acceleration voltage was 20 kV, spectra were taken in positive linear and reflectron modes, each mass spectrum was the average of 100–200 shots.

Equipment

MALDI-TOF sample preparation — Samples were dissolved in acetone or THF (4 mg/mL). DHB or 2,4,6-trihydroxy acetophenone were used as the matrices. They were dissolved in THF at a concentration of 10 mg/ml. Then 10 µl of each solution were taken and mixed. Finally, 2 µl of the mixed solution was pipetted onto the sample slide and air-dried.

Phenolic resin novolacs are produced by polycondensation of phenol or cresols and formaldehyde in the presence of an acidic catalyst. As a result, alternating copolymers are formed, where the phenolic units are bound in the polymer chain via methylene bridges. Since normally an excess of the phenol is present in the reaction mixture, the novolac chains have phenolic endgroups.

MALDI-TOF analysis

107 Da   106 Da   93 Da

The typical MALDI MS spectrum of a commercial novolac is presented in Fig. 6.35. The best spectra for this type of products are obtained using 2,4,6-trihydroxy acetophenone as the matrix and acetone as the solvent.

The spectrum consists of a number of equidistant peaks of different intensity. The peak-to-peak mass increment is 106 Da and exactly equals the mass of the repeat unit in the phenolic resin chain. This mass increment indicates that the sample is composed exclusively of phenol, and cresols are not present. Each peak rep-

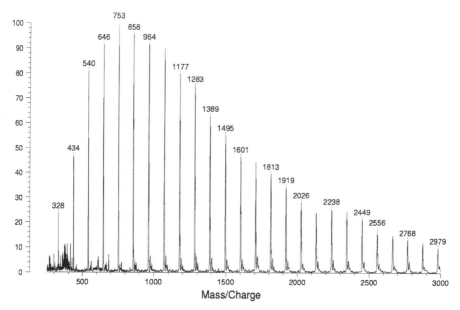

Fig. 6.35. MALDI-TOF spectrum of a phenolic novolac

resents one oligomer in the oligomer mixture. The peaks are due to the intact $[M+Na]^+$ molecular ions, which are formed by attachment of Na^+ from the matrix-analyte mixture to the oligomers. Accordingly, the masses of the oligomers correspond to $[M+Na]^+ = 223 + 106n$, n being the degree of polymerization and 223 g/mol being the mass of the endgroup and the attached Na^+ cation. From the masses at the peaks the corresponding degree of polymerization can be calculated, for example:

n = 13 $[M+Na]^+$ (calc.) 1601 g/mol
         $[M+Na]^+$ (exp.) 1601 g/mol

n = 15 $[M+Na]^+$ (calc.) 1813 g/mol
         $[M+Na]^+$ (exp.) 1813 g/mol

The spectrum shows that each oligomer peak is accompanied by a second peak of lower intensity. The mass increment for these pairs of peaks is 12–13 Da. The corresponding peak series can, in principle, be assigned to (A) linear oligomers with a quinone methide endgroup or (B) cyclic oligomers. Both structures are known to be present in novolacs:

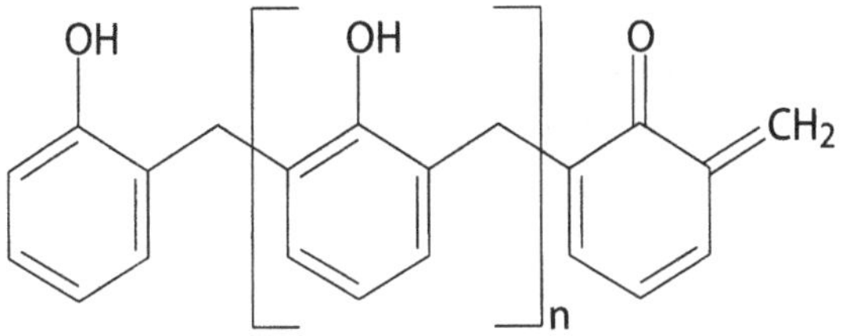

6.3 Analysis of Copolymers

[Structure A: novolac with fragments labeled 107 Da, 106 Da, 105 Da]

A

[Structure B: cyclic novolac with n+2 units]

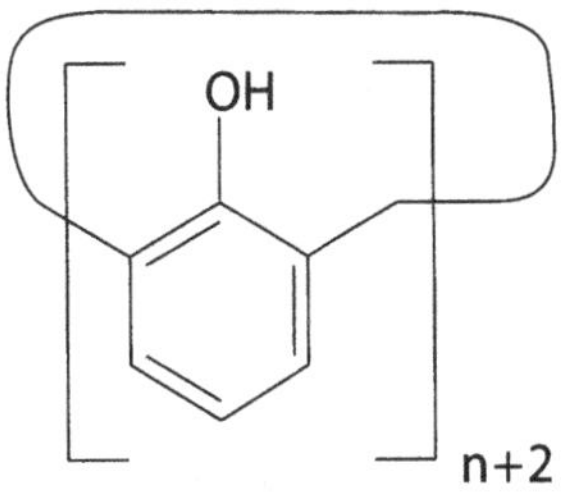

B

Unfortunately, both structures give the same peak masses and, therefore, cannot be differentiated by MALDI-TOF.

Since the molar masses of the novolacs are rather low, their molar mass distributions can be determined by MALDI-TOF by taking into account the intensity of the oligomer peaks and the mass at which they arise. As can be seen for two representative samples, the calculated data fit well with the nominal molar masses:

	Nominal	MALDI MS	
	M_w	M_w	M_n
novolac A	1800	1870	1390
novolac B	1400	1310	1080

Much more complicated than novolacs is the MALDI MS analysis of phenolic resols. This class of phenolic resins is prepared by base-catalyzed polycondensation of phenol and formaldehyde in a molar ratio of 1:1 to 1:3. As a result, polynuclear compounds having methylol groups bound to the phenolic nuclei, are formed:

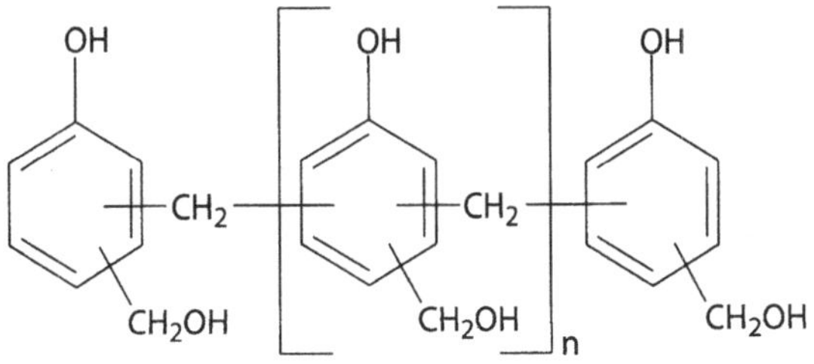

[Structure of phenolic resol with CH₂OH groups]

Depending on the molar ratio of phenol and formaldehyde, species with 1, 2, 3, or more methylol groups or mixtures of them are obtained, making the analysis of phenolic resols a difficult task. Using NMR spectroscopy, the average structure may be determined [54,55], the lower oligomers may be separated by HPLC [56,57]. A complete picture of all oligomers, however, has not been obtained so far.

The MALDI-TOF spectrum of a phenolic resol is presented in Fig. 6.36. Using 2,4,6-trihydroxy acetophenone as the matrix and THF as the solvent, well resolved peaks are obtained, corresponding to the intact $[M+Na]^+$ molecular ions of the respective oligomers. As the first reaction products dimethylol phenol (1) and trimethylol phenol (2) are formed, which produce mass peaks at 177 and 207 Da, respectively.

Peaks at 237 and 267 Da indicate that further formaldehyde is added to **2**, yielding hemiformal structures [58]:

1
M = 154 g/mol
$[M + Na]^+$ = 177 g/mol

2
M = 184 g/mol
$[M + Na]^+$ = 207 g/mol

Via condensation, the methylol phenols react with each other or with phenol, forming dinuclear compounds or higher oligomers, as is summarized in Table 6.13. The phenolic nuclei can be linked to each other via methylene or dimethylene ether bridges. Unfortunately, the mass of compound **3a** is equal to that of **3b**. Therefore, it is not possible to assign one single chemical structure to a peak in the mass spectrum:

$[M+Na]^+$ = 253 Da $[M+Na]^+$ = 253 Da
3a   3b

6.3 Analysis of Copolymers

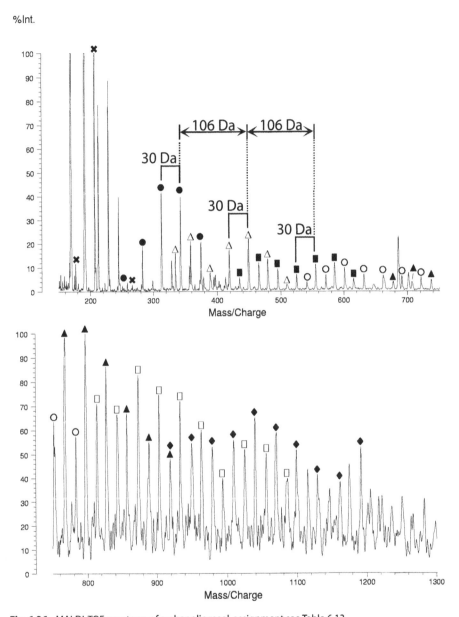

Fig. 6.36. MALDI-TOF spectrum of a phenolic resol, assignment see Table 6.13

Table 6.13. Assignment of the peaks $[M + Na]^+$ in the MALDI-TOF spectrum of a phenolic resol (P_XF_Z).

Series	Peak [Da]	X	Z	Series	Peak [Da]	X	Z
✖	177	1	2	▲	677	6	6
✖	207	1	3	▲	707	6	7
✖	237	1	4	▲	737	6	8
✖	267	1	5	▲	767	6	9
●	253	2	2	▲	797	6	10
●	283	2	3	▲	827	6	11
●	313	2	4	▲	857	6	12
●	343	2	5	▲	887	6	13
●	373	2	6	▲	917	6	14
△	329	3	2	☐	814	7	8
△	359	3	3	☐	844	7	9
△	389	3	4	☐	874	7	10
△	419	3	5	☐	904	7	11
△	449	3	6	☐	934	7	12
△	479	3	7	☐	964	7	13
△	509	3	8	☐	994	7	14
■	435	4	3	☐	1024	7	15
■	465	4	4	☐	1054	7	16
■	495	4	5	☐	1084	7	17
■	525	4	6	◆	919	8	9
■	555	4	7	◆	949	8	10
■	585	4	8	◆	979	8	11
■	615	4	9	◆	1009	8	12
○	541	5	4	◆	1039	8	13
○	571	5	5	◆	1069	8	14
○	601	5	6	◆	1099	8	15
○	631	5	7	◆	1129	8	16
○	661	5	8	◆	1159	8	17
○	691	5	9	◆	1189	8	18
○	721	5	10				
○	751	5	11				
○	781	5	12				

Since, in general, a phenolic resol may be regarded as a copolymer of the average structure P_xF_z, x being the number of phenolic nuclei P and z being the number of incorporated formaldehyde molecules F, each peak in the spectrum may be assigned to a certain P_xF_z. The complete assignment of all peaks in the spectrum is given in Table 6.13.

As was pointed out earlier, for specific applications phenolic resols are modified with urea or melamine. When a resin is produced from phenol, urea, and formaldehyde, different reaction

Table 6.14. Assignment of the peaks [M + Na]$^+$ for PF and UF condensates in the MALDI-TOF spectrum of a PUF resol ($P_XU_YF_Z$).

Series	Peak [Da]	X	Y	Z	Series	Peak [Da]	X	Y	Z
●	83	—	1	—	□	389	3	—	4
●	113	—	1	1	□	419	3	—	5
●	143	—	1	2	□	449	3	—	6
▲	155	—	2	1	□	479	3	—	7
▲	185	—	2	2	□	509	3	—	8
▲	215	—	2	3	◊	435	4	—	3
▲	245	—	2	4	◊	465	4	—	4
▲	275	—	2	5	◊	495	4	—	5
▲	305	—	2	6	◊	525	4	—	6
■	227	—	3	2	◊	555	4	—	7
■	257	—	3	3	◊	585	4	—	8
■	377	—	3	7	◊	615	4	—	9
■	407	—	3	8	▽	601	5	—	6
○	207	1	—	3	▽	631	5	—	7
○	237	1	—	4	▽	691	5	—	9
○	267	1	—	5	▽	721	5	—	10
△	253	2	—	2	▽	751	5	—	11
△	283	2	—	3	▽	781	5	—	12
△	313	2	—	4	◎	767	6	—	9
△	343	2	—	5	◎	797	6	—	10
△	373	2	—	6					

products can be formed, including phenol-formaldehyde, urea-formaldehyde, and phenol-urea-formaldehyde oligomers. These oligomers have different chemical compositions and different degrees of polymerization.

The MALDI-TOF spectrum of such a resin is presented in Fig. 6.37. As can be seen, the spectrum is very complex and it is complicated at first glance to make a proper peak assignment. However, the careful inspection of the spectrum reveals a peak order that is characteristic for PUF resins. Peak-to-peak mass increments of 106 Da and 72 Da are obtained that are typical for the repeating units of phenolic and urea resins, respectively. A mass increment of 30 Da indicates the attachment of an additional methylol group to a certain oligomer.

For example, the mass peaks at 207–237–267 Da and 253–283–313 Da are typical for phenol-formaldehyde oligomers, while the mass peaks at 83–113–143 Da and 155–185–215 Da indicate urea-formaldehyde oligomers:

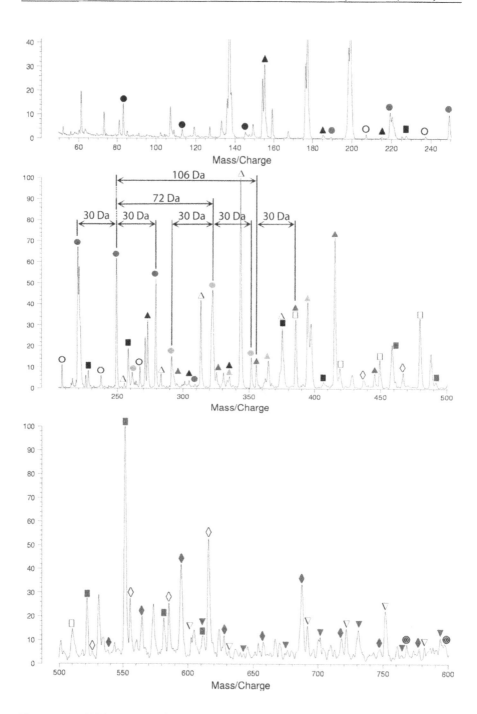

Fig. 6.37. MALDI-TOF spectrum of a urea-modified phenolic resol

[M+Na]⁺ = 83 Da   [M+Na]⁺ = 113 Da   [M+Na]⁺ = 143 Da

[M+Na]⁺ = 185 Da   [M+Na]⁺ = 215 Da

In addition to these types of structures, mixed phenol-urea co-condensates are formed which can be observed in the spectrum:

M+Na⁺ = 189 Da   M+Na⁺ = 219 Da

The assignments of the mass peaks in Fig. 6.37 are summarized in Tables 6.14 and 6.15.

6.3.6 Polyurethanes

Polyurethanes find widespread use as materials for foams, shoe soles, steering wheels, and others. These elastomeric materials are multiblock copolymers with an alternating sequence of hard and soft blocks. The soft blocks usually consist of multifunctional polyether or polyester polyols. By reaction of diisocyanates with the polyol (and a short chain diol as chain extender) urethane linkages are formed. The chain extender and the diisocyanates form the hard phase. As hard and soft segments are incompatible, microphase separation occurs upon cooling. For establishing structure-property relationships the molecular constitution is of interest especially as the macroscopic mechanical behavior can be

Aim

Table 6.15. Assignment of the peaks [M + Na]$^+$ for PUF co-condensates in the MALDI-TOF spectrum of a PUF resol ($P_XU_yF_z$)

Series	Peak [Da]	X	Y	Z	Series	Peak [Da]	X	Y	Z
●	189	1	1	1	◆	537	4	1	5
●	219	1	1	2	◆	567	4	1	6
●	249	1	1	3	◆	597	4	1	7
●	279	1	1	4	◆	627	4	1	8
●	309	1	1	5	◆	657	4	1	9
▲	295	2	1	2	◆	687	4	1	10
▲	325	2	1	3	◆	717	4	1	11
▲	355	2	1	4	◆	747	4	1	12
▲	385	2	1	5	◆	777	4	1	13
▲	415	2	1	6	▼	613	5	1	5
▲	445	2	1	7	▼	643	5	1	6
■	461	3	1	5	▼	673	5	1	7
■	491	3	1	6	▼	703	5	1	8
■	521	3	1	7	▼	733	5	1	9
■	551	3	1	8	▼	763	5	1	10
■	581	3	1	9	▼	793	5	1	11
■	611	3	1	10					
⬤	261	1	2	2	▲	333	1	3	3
⬤	291	1	2	3	▲	363	1	3	4
⬤	321	1	2	4	▲	393	1	3	5
⬤	351	1	2	5					

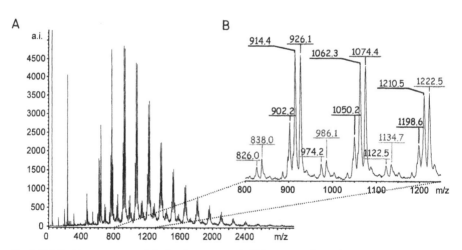

Fig. 6.38 A,B. Mass spectrum of the hard segments of a TDI-polyester foam: **A** overview spectrum; **B** spectral details

6.3 Analysis of Copolymers

Table 6.16. Signal assignment for the peaks in Fig. 6.38B (a = amine endgroup; d = dipropylene glycol endgroup; e = ethanol endgroup; t = TDI repeat unit (148.1 Da))

Peak mass (Da)	Assignment
902.2	a-6t-a
914.1	a-5t-d
926.1	d-4t-d
975.2	a-6t-e
986.1	d-5t-e
1050.2	a-7t-a

tuned over a broad range by the specific compositions. Using MALDI MS the molecular constitution can be investigated for the hard [75,76] and soft phase [73].

To investigate the hard segments a selective degradation technique [76] was applied to cleave the soft segments while leaving the hard segments unchanged: alkaline hydrolysis of the crushed samples for 12h in a 5 wt% solution of NaOH in an ethanol/water mixture (1/1 v/v). This mixture is then diluted with water, the remaining hard phase isolated by centrifugation, washed and dried.

Polyurethane foam prepared from toluenediisocyanate (TDI) and adipic acid-dipropylene glycol polyester (BASF Aktiengesellschaft, Ludwigshafen) after hydrolysis to isolate the hard segments. — Materials

MALDI-TOF system — Bruker Biflex, accelerating voltage 30 kV, 100 laser shots were accumulated in the linear mode. — Equipment

MALDI-TOF sample preparation — the hard phase and the matrix dithranol were dissolved in DMSO (10 resp. 20 mg/ml), cationizing agent was a 0.1 molar solution of K-TFA. Then 1 µl of a 10:10:1 mixture was spotted on the steel target and air-dried.

A typical MALDI-TOF mass spectrum of the hard phase of a TDI-polyester foam is shown in Fig. 6.38 displaying an oligomer distribution up to 2800 Da. The assignment of the signals in Fig. 6.38B is given in Table 6.16. — MALDI-TOF analysis

The polyester in this case simply consists of adipic acid and dipropylene glycol. The series of the main peaks consists of triplets. The first signal at 902 Da results from a hard segment with six TDI monomers and amine groups at both ends:

a-6t-a

$[M+K]^+ = 39 (K^+) + 16$ (endgroup) $+ 106$ (endgroup) $+ 5 \times 148.1 = 902$ Da

The second one originates from a segment with 5 TDI units with an amine group and a dipropylene glycol urethane at the ends. The third peak, again 12 Da higher, can be explained as a molecule with four TDI units and two dipropylene glycols at the ends. By control experiments it could be confirmed that the amine groups already did exist in the foam and are not a consequence of the degradation process.

Besides the smaller triplet due to Na^+-impurities two additional signals appear, for example at 974 and 986 Da, which are caused by transesterification during the hydrolysis: a dipropylene glycol moiety is substituted by an ethanol unit.

In conclusion the MALDI-TOF results not only allow the determination of the oligomer distribution of the hard segments but also to elucidate their connection to the soft phase.

An extended analysis [76] demonstrated that the length of the hard segments decreases slower for polyester than for polyether foams. This finding was attributed to a lower solubility of the hard segments in polyethers, which leads to an earlier onset of phase separation. So both mobility and reactivity in the segregated hard domains are reduced. A finding in agreement with results of Yontz and Hsu [75] is the fact that the hard segment oligomer distribution is rather independent of the water content in the initial formulation.

Mehl et al. [73] on the contrary investigated the soft blocks of polyether and polyester urethanes. Here the soft phases were recovered by degradation of the polyether urethane in ethanolamine and of the polyester urethane in phenylisocyanate. The MALDI-TOF analysis revealed apart from minor changes the original polyol distribution. Partial degradation allows the identification of the diisocyanate component by analyzing ions that contain the relevant linkage.

In summary it can be stated that MALDI-TOF provides a proper means for molecular characterization of complex materials like polyurethane systems.

6.3.7 Further Applications

Further applications on the analysis of copolymers by MALDI-TOF mass spectrometry are presented in Table 6.17.

Table 6.17. Applications of MALDI-TOF MS for analysis of copolymers

Polymer	Matrix	Remarks	Reference
Polyhydroxy alkanoates	DHB	Analysis of partially transesterified polyhydroxy alkanoate (PHA) produced from bacteria using saponified vegetable oils as the carbon source, PHA oligomers up to 10,000 Da were detected	[59]
Polycarbonate	HABA	Investigation of the pyrolysis of polycarbonate at different temperatures, elucidation of the degradation mechanism	[60]
Polyglycerols	α-Cyano HCA	Synthesis and analysis of hyperbranched polyglycerols, ring-opening multibranching polymerization, degree of polymerization 5 to 83	[61]
Aliphatic polyesters	HABA	Analysis of biodegradable polyesters, succinate/adipate and succinate/sebacate co-polyesters, determination of cyclic oligomers, calibration of SEC	[62]
Polysiloxanes	2-NPOE	Polydimethylsiloxane-co-poly(hydromethylsiloxane), peak assignment through computing the expected MALDI spectrum from SEC and NMR data	[63]
Polydimethylsiloxanes	Dithranol	Comparison of SEC and MALDI-TOF data, MALDI underestimates the higher molar mass fractions, formation of different structures due to interchain-exchange mechanisms	[64]
Polybutadiene star polymers	Dithranol	Analysis of structural quality concerning arm number and polydispersity	[65]
Polyacetylene macrocycles	Dithranol	Analysis of shape persistent macrocycles of poly(phenylene-acetylene)	[66]
Copolymers based on PEO and PIB	α-Cyano HCA	Synthesis and analysis of amphiphilic diblock and multiarm copolymers, MALDI-TOF reveals number of arms	[67]
PMMA-based block copolymers	Dithranol	Block copolymers by endgroup reaction of functional PMMA, analysis of PMMA-b-PEO, PMMA-b-polymethacrylic acid, PMMA-b-PDMS-b-PMMA	[68]

Table 6.17. (continued)

Polymer	Matrix	Remarks	Reference
PMMA-based block copolymers	Dithranol	Block copolymers by endgroup reaction of functional PMMA, analysis of PMMA-*b*-PEO, PMMA-*b*-polymethacrylic acid, PMMA-*b*-PDMS-*b*-PMMA	[68]
PS-b-PI block copolymers	Dithranol	Synthesis and analysis of bifunctional poly (styrene-*b*-isoprene) by anionic polymerization	[69]
Co-oligolactones	Dithranol	Copolymerization of *L*-lactide and δ-valerolactone or ε-caprolactone	[70]
Epoxy resins	DHB	Alternating ionic polymerization of phenyl glycidyl ether and phthalic anhydride, analysis of the mechanism	[71]
Vinylhexaphenylbenzene copolymers	Dithranol	Synthesis and analysis of copolymers of vinylhexaphenylbenzene with MMA and styrene	[72]
Polyurethanes	Dithranol	Analysis of polyether and polyester polyurethane soft blocks after selective degradation of the PU	[73]
Copolyformals	IAA	Polycondensation of fullerene containing monomers, detection of linears and cyclics	[74]

References

1. RÄDER HJ, SCHREPP W (1998) Acta Polym 49:272
2. PASCH H, GHAHARY R (2000) Macromol Symp 152:267
3. ENTELIS SG, EVREINOV VV, GORSHKOV AV (1986) Adv Polym Sci 76:129
4. ENTELIS SG, EVREINOV VV, KUZAEV AI (1985) Reactive oligomers. Khimiya Publishers, Moscow
5. MONTAUDO G, MONTAUDO MS, PUGLISI C, SAMPERI F (1996) J Polym Sci Polym Chem A34:439
6. BRAUN D, GHAHARY R, PASCH H (1996) Polymer 37:777
7. BRAUN D, GHAHARY R, ZISER T (1995) Angew Makromol Chem 233:121
8. MONTAUDO G, MONTAUDO MS, PUGLISI C, SAMPERI F, SEPULCHRE M (1996) Macromol Chem Phys 197:2615
9. MATYJASZEWSKI K, GRESZTA D, HRKACH JS, KIM HK (1995) Macromolecules 28:59
10. PASCH H, GORES F (1995) Polymer 36:1999

11. PRODUCTS OF POLYMER STANDARDS SERVICE, Mainz, Germany
12. DANIS PO, KARR DE, SIMONSICK WJ, WU DT (1995) Macromolecules 28:1229
13. KOWALSKI A, DUDA A, PENCZEK S (2000) Macromolecules 33:689
14. DOI Y, LEMSTRA PJ, NIJENHUIS AJ, VAN AERT HAM, BASTIAANSEN C (1995) Macromolecules 28:2124
15. KRICHELDORF HR, KREISER-SAUNDERS I, BOETTCHER C (1995) Polymer 36:1253
16. ZHANG X, WYSS UP, PICHORA D, GOOSEN MFA (1992) Polym Bull 27:623
17. KOWALSKI A, DUDA A, PENCZEK S (1998) Macromol Rapid Commun 19:567
18. PENCZEK S, DUDA A, KOWALSKI A, LIBISZOWSKI J (1999) Polym Mat Sci Eng 80:95
19. PASCH H, UNVERICHT R, RESCH M (1993) Angew Makromol Chem 212:191
20. HUNT KH, CROSSMAN MC, HADDLETON DM, LLOYD PM, DERRICK PJ (1995) Macromol Rapid Commun 16:725
21. SPICKERMANN J, RÄDER HJ, MÜLLEN K, MÜLLER B (1996) Macromol Rapid Commun 17:885
22. THOMSON B, SUDDABY K, RUDIN A, LAJOIE G (1996) Eur Polym J 32:239
23. WHITTAL RM, LI L, LEE S, WINNIK MA (1996) Macromol Rapid Commun 17:59
24. WILLIAMS JB, GUSEV AI, HERCULES DM (1997) Macromolecules 30:3781
25. ZAMMIT MD, DAVIS TP, HADDLETON DM, SUDDABY KG (1997) Macromolecules 30:1915
26. NAGASAKI Y, OGAWA R, YAMAMOTO S, KATO M, KATAOKA K (1997) Macromolecules 30:6489
27. HAMILTON SC, SEMLYEN JA, HADDLETON DM (1998) Polymer 39:3241
28. KARPFENSTEIN HM, DAVIS TP (1998) Macromol Chem Phys 199:2403
29. MAHON A, KEMP TJ, VARNEY JE, DERRICK PJ (1998) Polymer 39:6213
30. DOURGES MA, CHARLEUX B, VAIRON JP, BLAIS JC, BOLBACH G, TABET JC (1999) Macromolecules 32:2495
31. LIU J, LOEWE RS, MCCULLOUGH RD (1999) Macromolecules 32:5777
32. AYORINDE FO, ERIBO BE, JOHNSON JH, ELHILO E (1999) Rapid Commun Mass Spectrom 13:1124
33. PUGLISI C, SAMPERI F, CARROCCIO S, MONTAUDO G (1999) Rapid Commun Mass Spectrom 13:2260

34. Puglisi C, Samperi F, Carroccio S, Montaudo G (1999) Rapid Commun Mass Spectrom 13:2268
35. Völcker NH, Klee D, Hanna M, Höcker H, Bou JJ, de Ilarduya AM, Munoz-Guerra S (1999) Macromol Chem Phys 200:1363
36. Kowalski A, Duda A, Penczek S (2000) Macromolecules 33:7359
37. Keki S, Deak G, Mayer-Posner FJ, Zsuga M (2000) Macromol Rapid Commun 21:770
38. Shen X, He X, Chen G, Zhou P, Huang L (2000) Macromol Rapid Commun 21:1162
39. Rokicki G, Kowalczyk T (2000) Polymer 41:9013
40. Nanda AK, Ganesh K, Kishore K, Surinarayanan (2000) Polymer 41:9063
41. Sato H, Ohtani H, Tsuge S, Hayashi N, Katoh K, Masuda E, Ohnishi K (2001) Rapid Commun Mass Spectrom 15:82
42. Weidner S, Kühn G, Just U (1995) Rapid Commun Mass Spectrom 9:697
43. Weidner S, Kühn G, Friedrich J, Schröder H (1996) Rapid Commun Mass Spectrom 10:40
44. Weidner S, Kühn G, Werthmann B, Schröder H, Just U, Borowski R, Decker R, Schwarz B, Schmücking I, Seifert I (1997) J Polym Sci Polym Chem 35:2183
45. Wilczek-Vera G, Yu Y, Waddell K, Danis PO, Eisenberg A (1999) Rapid Commun Mass Spectrom 13:764
46. Wilczek-Vera G, Danis PO, Eisenberg A (1996) Macromolecules 29:4036
47. Schriemer DC, Whittal RM, Li L (1997) Macromolecules 30:1955
48. Whittal RM, Li L (1995) Anal Chem 67:1950
49. Chen R, Zhang N, Tseng AM, Li L (2000) Rapid Commun Mass Spectrom 14:2175
50. Yoshida S, Yamamoto S, Takamatsu T (1998) Rapid Commun Mass Spectrom 12:535
51. Montaudo G, Garozzo D, Montaudo MS, Puglisi C, Samperi F (1995) Macromolecules 28:7983
52. Räder HJ, Spickermann J, Müllen K (1995) Macromol Chem Phys 196:3967
53. Pasch H, Rode K, Ghahary R, Braun D (1996) Angew Makromol Chem 241:95
54. Woodbrey CJ, Higginbottom HP, Culbertson HM (1965) J Polym Sci A 3:1079
55. Pasch H, Goetzky P, Gründemann E, Raubach H (1981) Acta Polym 32:14

56. VAN DER MAEDEN FPB, BIEMOND MEF, JANSSEN PC (1978) J Chromatogr 149:539
57. MUCH H, PASCH H (1982) Acta Polym 33:366
58. PASCH H, GOETZKY P, RAUBACH H (1983) Acta Polym 34:150
59. SAEED KA, AYORINDE FO, ERIBO BE, GORDON M, COLLIER L (1999) Rapid Commun Mass Spectrom 13:1951
60. PUGLISI C, SAMPERI F, CARROCCIO S, MONTAUDO G (1999) Macromolecules 32:8821
61. SUNDER A, HANSELMANN R, FREY H, MÜLHAUPT R (1999) Macromolecules 32:4240
62. CARROCCIO S, RIZZARELLI P, PUGLISI C (2000) Rapid Commun Mass Spectrom 14:1513
63. SERVATY S, KÖHLER W, MEYER WH, ROSENAUER C, SPICKERMANN J, RÄDER HJ, WEGNER G (1998) Macromolecules 31:2468
64. HADDLETON DM, BON SA, ROBINSON KL, EMERY NJ, MOSS I (2000) Macromol Chem Phys 201:694
65. ALLGAIER J, MARTIN K, RÄDER HJ, MÜLLEN K (1999) Macromolecules 32:3190
66. HÖGER S, SPICKERMANN J, MORRISON DL, DZIEZOK P, RÄDER HJ (1997) Macromolecules 30:3110
67. LEMAIRE C, TESSIER M, MARECHAL E (1997) Macromol Symp 122:371
68. ESSELBORN E, FOCK J, KNEBELKAMP A (1996) Macromol Symp 102:91
69. SCHÄDLER V, SPICKERMANN J, RÄDER HJ, WIESNER U (1996) Macromolecules 29:4865
70. GOPP U, SANDNER B, SCHÖCH M (1998) Macromol Symp 130:113
71. LEUKEL J, BURCHARD W, KRÜGER RP, MUCH H, SCHULZ G (1996) Macromol Rapid Commun 17:359
72. KÜBEL C, CHEN SL, MÜLLEN K (1998) Macromolecules 31:6014
73. MEHL JT, MURGASOVA R, DONG X, HERCULES DM (2000) Anal Chem 72:2490
74. SCAMPORRINO E, VITALINI D, MINEO P (1999) Macromolecules 32:4247
75. YONTZ DJ, HSU SL (2000) Macromolecules 33:8415
76. MERTES J, STUTZ H, SCHREPP W, KREYENSCHMIDT M (1998) J Cell Plast 34:526

7 Coupling of Liquid Chromatography and MALDI-TOF Mass Spectrometry

7.1 Introduction

From the very early stages of development of modern mass spectrometry, the value of its combination with chromatography was quickly recognized. The coupling of GC with MS was a natural evolution since they are both vapor phase techniques, and very quickly GC-MS has been accepted as a standard component of the organic analytical laboratory. It has taken considerably longer to achieve a satisfactory and all-purpose mode of HPLC-MS coupling. The difficulties with HPLC-MS were associated with the fact that vaporization of typically 1 ml/min from the HPLC translates into a vapor flow rate of approximately 500-1000 ml/min. Other difficulties related to the eluent composition as result of the frequent use of non-volatile modifiers, and the ionization of non-volatile and thermally labile analytes. However, during the past few years commercial interfaces have been developed which have led to a broad applicability of HPLC-MS [1-3].

The techniques necessary for the successful introduction of a liquid stream into a mass spectrometer are based on the following principles: electrospray ionization [4], atmospheric pressure chemical ionization [5], thermospray ionization [6], and particle beam ionization [7]. In a thermospray interface [8-10] a jet of vapor and small droplets is generated out of a heated vaporizer tube. Nebulization takes place as a result of the disruption of the liquid by the expanding vapor that is formed at the tube wall upon evaporation of part of the liquid in the tube. Prior to the onset of the partial evaporation inside the tube a considerable amount of heat is transferred to the solvent. This heat later on assists in the desolvation of the droplets in the low-pressure region. By applying efficient pumping by means of a high-throughput mechanical pump attached directly to the ion source up to 2 ml/min of aqueous solvents can be introduced into the MS vacuum system. The ionization of the analytes takes place by means of solvent-mediated chemical ionization reactions and ion evaporation processes.

In a particle beam interface [7,11,12] the column effluent is nebulized either pneumatically or by thermospray nebulization, into a near atmospheric-pressure desolvation chamber, which is connected to a momentum separator where the high-mass analytes are preferentially transferred to the MS ion source while the low-mass solvent molecules are efficiently pumped away. The analyte molecules are transferred as small particles to a conventional ion source, where they disintegrate upon collisions at the heated source walls. The released gaseous molecules are ionized by EI or CI.

Two different sample-introduction approaches are used in combination with atmospheric pressure ionization (API) devices. They primarily differ in the nebulization principle and in the application range they cover. In a heated nebulizer or APCI interface [13], the column effluent is pneumatically nebulized in a heated tube, where the solvent evaporation is almost completed. Atmospheric-pressure chemical ionization (APCI), initiated by electrons from a corona discharge needle, is achieved in the same region. Subsequently, the ions generated are sampled into the high vacuum of the mass spectrometer for mass analysis.

In an electrospray interface [14,15], the column effluent is nebulized into the atmospheric-pressure region as a result of the action of a high electric field resulting from a 3-kV potential difference between the narrow-bore spray capillary and a surrounding counter electrode. The solvent emerging from the capillary breaks into fine threads which subsequently disintegrate in small droplets. In some designs, the electrospray nebulization is assisted by pneumatic nebulization. Such an approach is called an ionspray interface [16]. For more details on electrospray ionization see Chap. 2 and [90].

7.2 Need for LC-MALDI-TOF Coupling

Typical operating parameters of the various LC-MS interfaces, especially in terms of LC conditions, are summarized in Table 7.1 [3]. They indicate that only a very limited number of the existing interfaces can directly be coupled to typical LC equipment running on flow rates of 0.5 to 1 ml/min. In addition, some of them are best used with aqueous mobile phases and buffers.

From the point of view of polymer analysis, a mass spectrometric detector would be a most interesting alternative to the conventional detectors because this detector could provide absolute molar masses of polymer components [17,18]. Provided that fragmentation does not occur, intact molecular ions could be measured. The measured mass of a particular component could then be correlated with chemical composition or chain length.

7.2 Need for LC-MALDI-TOF Coupling

Table 7.1. Comparison of LC-MS interfaces in terms of LC conditions

Interface	Maximum flow rate (ml/min)	Mobile phase composition
Direct liquid introduction	0.05	Reversed-phase (RP) solvents, no buffers
Thermospray	2	RP solvents with volatile buffers
Particle-beam	0.5	RP solvents with volatile buffers
Electrospray	0.5	RP solvents with volatile buffers
Heated nebulizer/APCI	2	RP solvents with volatile buffers

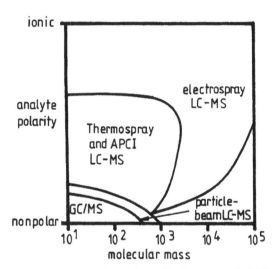

Fig. 7.1. Representation of the application ranges of LC-MS interfaces. (Reprinted from [3] with permission of Elsevier Science B.V., The Netherlands)

However, the major drawback of most conventional HPLC-MS techniques is the limited mass range, preventing higher oligomers (molar mass above 2000–3000 g/mol) from being ionized without fragmentation [19–21]. A schematic diagram, indicating the application areas of the various LC-MS interfaces in terms of analyte polarity and molar mass is given in Fig. 7.1.

As can be seen, there is only electrospray LC-MS covering a larger molar mass range. Accordingly, out of all existing LC-MS techniques, electrospray ionization received the most attention in polymer analysis. It has been widely applied in biopolymer analysis. Proteins and biopolymers are typically ionized through acid-base equilibria. When a protein solution (the effluent from an HPLC separation) is exposed to an electrical potential it ionizes and disperses into charged droplets. Larger molecules aquire more than one single charge, and, typically, a mixture of differently charged ions is obtained.

Unfortunately, most synthetic polymers have no acidic or basic functional groups that can be used for ion formation. Moreover, each molecule gives rise to a charge distribution envelope, thus further complicating the spectrum. Therefore, synthetic polymers that can typically contain a distribution of chain lengths and a variety of chemical composition or functionality furnish complicated mass spectra, making interpretation nearly impossible.

Since no other promising LC-MS technique has been available, different groups tried to overcome the limitations of ESI-MS. Simonsick and Prokai added sodium cations to the mobile phase to facilitate ionization [22,23]. To simplify the resulting ESI spectra, the number of components entering the ion source was reduced by a chromatographic prefractionation. For example, by combining SEC with electrospray detection, the elution curves of polyethylene oxides could be calibrated through ESI-MS analysis of the SEC fractions. In another application, the chemical composition distribution of acrylic macromonomers was profiled across the molar mass distribution. The analysis of polyethylene oxides by SEC-ESI-MS with respect to chemical composition and oligomer distribution was discussed by Simonsick [24]. In a similar approach aliphatic polyesters [25], phenolic resins [26], methyl methacrylate macromonomers [26], and polysulfides have been analyzed [27]. The detectable mass range for different species, however, was well below 5000 g/mol, indicating that the technique is not really suited for high molar mass polymer analysis.

Considering the potential of MALDI-TOF in terms of versatility and sensitivity, on-line coupling with liquid chromatography would be a highly attractive possibility. In the analysis of synthetic polymers, the great promise of MALDI-TOF MS is to perform the direct identification of mass-resolved polymer chains, including intact oligomers within a molar mass distribution, and the simultaneous determination of structure and endgroups in polymer samples. Compared to other mass spectrometric techniques, the accessible mass range is virtually unlimited, mostly singly-charged ions are obtained, and the technique is fast and instrumentally very simple. Moreover, relatively inexpensive commercial instrumentation has become accessible.

Unfortunately, MALDI-TOF is based on the desorption of molecules from a solid surface layer and, therefore, a priori not compatible with liquid chromatography. In an attempt to take advantage of the MALDI-TOF capabilities, a number of research groups carried out off-line LC separations and subjected the resulting fractions to MALDI-TOF measurements. Although this is laborious, it has the advantage that virtually any type of chromatographic separation can be combined with MALDI-TOF.

7.3 Off-line Measurement of LC Fractions

The different options for using MALDI-TOF MS as an off-line detector in liquid chromatography have been discussed by Nielen [100], Gusev [101], and Pasch and Rode [28]. In SEC of low molar mass samples the separation into individual oligomers and the quantitative determination of the molar mass distribution via an oligomer calibration could be achieved; see for example Fig. 7.2A for oligo(caprolactone). The lower oligomers appeared as well separated peaks at the high retention time end of the chromatogram. For the analysis of the peaks, i.e., the assignment of a certain degree of polymerization (n) to each peak, MALDI-TOF MS was used. The SEC separation was conducted at the usual analytical scale and the oligomer fractions were collected, resulting in amounts of 5–20 ng substance per fraction in THF solution. The solutions were directly mixed with the matrix solution, placed on the sample slide and subjected to the MALDI experiments. As a large number of fractions may be introduced into the mass spectrometer at one time, sample preparation and MALDI MS measurements take a very short period of time. In total nine fractions were collected from SEC and measured by MALDI MS. For the lower oligomers the spectra consisted of a number of peaks of high intensity, having a peak-to-peak mass increment of 114 Da, which equals the mass of the caprolactone repeating unit. These peaks represented the $[M+Na]^+$ molecular ions, whereas the peaks of lower intensity in their vicinity were due to the formation of $[M+K]^+$ molecular ions. $[M+Na]^+$ and $[M+K]^+$ molecular ions were formed due to the presence of small amounts of Na^+ and K^+ ions in the samples and/or the matrix. Further peaks of low intensity indicated a functional heterogeneity in the samples. From the masses of the $[M+Na]^+$ peaks the degree of polymerization of the corresponding oligomer was calculated. By this procedure, the first peak in the chromatogram was assigned to $n=1$, the second peak to $n=2$, and so on. From the elution time and the degree of polymerization of each oligomer peak an oligomer calibration curve log molar mass vs elution time was constructed. The conventional calibration curve based on polystyrene standards differed remarkably from the oligomer calibration curve, as can be seen in Fig. 7.2B. In the following sections a number of typical applications for off-line coupling of liquid

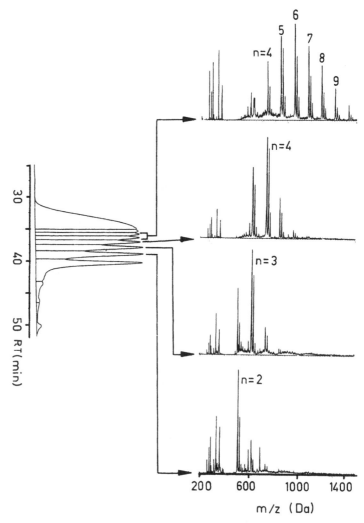

Fig. 7.2. A SEC of an oligo(caprolactone) and off-line analysis of fractions by MALDI-TOF, peak assignment indicates degree of polymerization (n)

chromatography and MALDI-TOF MS shall be described in detail.

7.3.1 Calibration of Size Exclusion Chromatography [29]

Aim

One of the limitations of size exclusion chromatography is that it is a relative method. Accordingly, for accurate molar mass determination of a certain polymer sample, the SEC system must be calibrated using narrow-disperse standards of the same chemical composition. In many cases, especially for copolymers, such cali-

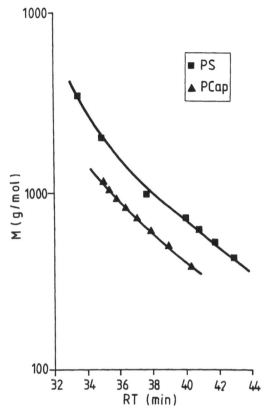

Fig. 7.2. B calibration graphs of molar mass vs retention time for (■) polystyrene and (▲) poly(caprolactone). (Reprinted from [28] with permission of Elsevier Science B.V., The Netherlands)

bration standards are not available and the calibration is carried out with available standards of other chemical compositions, very frequently polystyrene or polymethyl methacrylate. Such an approach, however, is a source of error since the hydrodynamic properties of the samples under investigation do not fit the hydrodynamic properties of the calibrants.

To overcome the calibration dilemma MALDI-TOF MS can be used. By carrying out a SEC separation under semiprep conditions, narrow disperse SEC fractions can be accumulated that can be analyzed easily by MALDI-TOF afterwords.

Commercial polymethyl methacrylate (PMMA, 25,000 g/mol), polyvinyl acetate (PVAc, 40,000 g/mol), and a copolymer of vinyl acetate and vinyl pyrrolidone (PNVP/PVAc, 40,000 g/mol) **Materials**

Chromatographic System — modular SEC system comprising of a Waters Model 6000A pump and a Waters Model 401 differential refractometer. Between the column and the detector the flow **Equipment** was split with 60% going to the detector and 40% being collected in vials for later MALDI-TOF analysis.

Columns — Polymer Labs PLgel MiniMIX, average particle size 3 μm, column size 25×4 mm.

Mobile phase — tetrahydrofuran.

Sample amount — 20 μl of a 10-20 g/l solution in THF.

MALDI-TOF system — Bruker Reflex MALDI-TOF, acceleration voltage of positive 30 kV.

MALDI-TOF sample preparation — To the dried sample fractions 3 mL of the matrix solution (0.2 mol/l *trans*-3-indoleacrylic acid in acetone) was added. An aliquot of this volume was dropped on the MALDI sample target, dried, and analyzed.

Preparatory investigations

As a representative example, first PMMA is fractionated by size exclusion chromatography; see Fig. 7.3A. The total elution volume for this fractionation is about 3 ml, in total six fractions are collected after different elution volumes. These fractions are analyzed subsequently by MALDI-TOF mass spectrometry, the corresponding MALDI-TOF spectra for the fractions collected after 1.65, 1.74, 1.91, and 2.09 ml are presented in Fig. 7.3B. As can be seen very clearly, different elution volumes correspond to different molar masses of the fractions. An important feature of these spectra is that since the collection time is short relative to the rate of elution of the polymer, then the shape of the distributions in Fig. 7.3B reflects the bandwidth of the SEC system. This function can then be used for deconvolution of the measured chromatogram to give the true shape of the distribution.

MALDI-TOF analysis

From the mass spectra of each individual fraction the peak of the mass distribution can be determined and plotted vs the elution volume. The plot log molar mass vs elution volume is then the specific SEC calibration function for PMMA. In the present case a linear relationship is obtained (R = 0.9994); see Fig. 7.4. This relationship is used subsequently for calibration of the SEC curve in Fig. 7.3A and calculation of the molar mass distribution. For the present sample the following numbers are calculated: $M_n = 15,000$, $M_w = 26,600$, $M_w/M_n = 1.79$. This agrees well with the given value of the M_w of 26,600 g/mol.

Similar measurements are made for PVAc and PNVP/VAc and their molar mass vs elution volume relationships are plotted in Fig. 7.4 as well. There is an obvious difference in slope and position for the calibration lines, indicating that each polymer behaves differently and their hydrodynamic volumes scale with mass in unique fashions.

One additional feature of the MALDI-TOF analysis of SEC fractions is that at lower m/z values significant compositional information can be obtained. As is shown in Fig. 7.5 for a fraction of

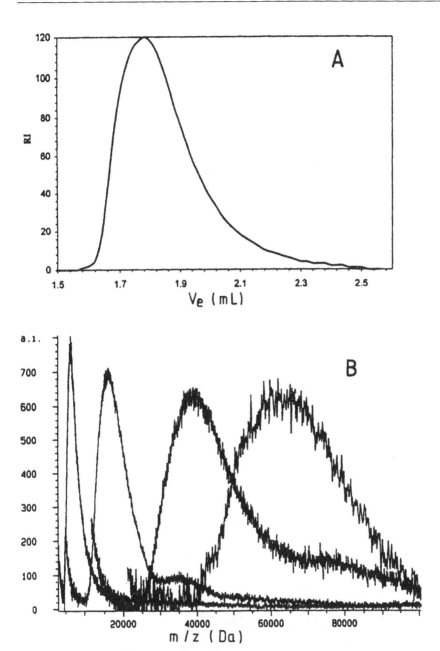

Fig. 7.3. **A** SEC chromatogram for PMMA. **B** MALDI-TOF spectra of PMMA fractions collected at elution volumes of 2.09, 1.91, 1.74, and 1.65 ml (*left to right*). (Reprinted from [29] with permission of American Chemical Society, U.S.A.)

PNVP/VAc, individual oligomers can be separated. From such a spectrum the endgroup, monomer composition, and monomer ratio can be determined [30].

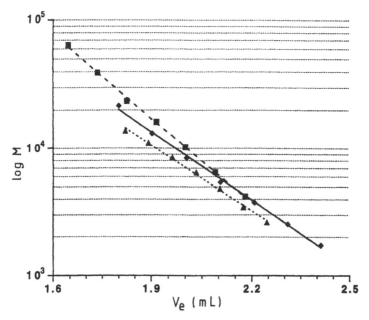

Fig. 7.4. Plots of log molar mass vs retention volume for PMMA (■), PVAc (▲), and PNVP/VAc (◆). (Reprinted from [29] with permission of American Chemical Society, U.S.A.)

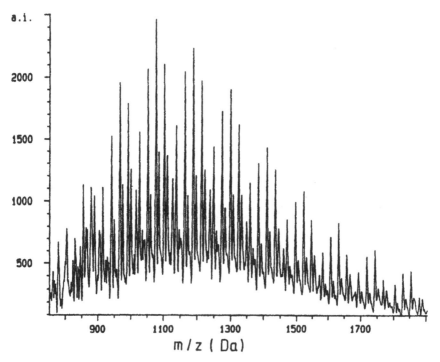

Fig. 7.5. MALDI-TOF spectrum of the 2.41-ml fraction of PNVP/VAc measured in reflectron mode. (Reprinted from [29] with permission of American Chemical Society, U.S.A.)

7.3.2 Molar Mass Distribution of Polyester Copolymers [31,32]

Polyesters are polycondensation products with a broad molar mass distribution. The determination of MMD by SEC is complicated due to the fact that calibration standards with narrow polydispersity are not commercially available. The direct MMD determination by MALDI-TOF mass spectrometry is not possible because the expected polydispersity index of $M_w/M_n \sim 2$ is too high for direct measurement. Even more complicated is the situation for polyester copolymers where different diols and diacids are reacted with each other. In this case an accurate molar mass determination by SEC is not possible unless molar mass sensitive detectors (light scattering, viscometer detectors) are coupled to SEC. — Aim

By prefractionating different polyester samples, fractions of low polydispersity can be obtained which subsequently can be analyzed by MALDI-TOF MS. These analyzes yield calibration curves for polyesters of different composition which can be used for computing molar mass averages for homo- and copolymers.

Laboratory products of different polyesters have been prepared by melt polycondensation starting from stoichiometric amounts of dimethyl esters and 1,4-butane diol. As the dimethyl esters the esters of the following acids were used: adipic acid, succinic acid, and sebacic acid. — Materials

Chromatographic system — Waters Model 600A apparatus and a Waters Model 401 differential refractometer. Thirty drops of each fraction were collected, fractionation was carried out several times to accumulate sufficient amounts for further analyzes. — Equipment

Columns — five Ultrastyragel connected in series, average particle size 10 µm, column size 300×7.8 mm, pore sizes 10^5, 10^3, 500, 10^4, 100 Å.

Mobile phase — tetrahydrofuran or chloroform.

Sample amount — 60 µl of a 15 g/L solution in THF or chloroform.

MALDI-TOF System — Bruker Reflex MALDI-TOF with linear and reflectron detectors, acceleration voltage of positive 30 kV.

MALDI-TOF sample preparation — 0.1 ml of each collected fraction was mixed with 2 ml HABA solution (0.1 mol/l 2-(4-hydroxyphenyl azo)benzoic acid in THF/CHCl$_3$ 1:1). Then 0.2 µl of this volume were dropped on the MALDI sample target, dried, and analyzed.

As the first step of these investigations, the homopolymers polybutylene adipate (PBA), polybutylene succinate (PBSu) and polybutylene sebacate (PBSe) are fractionated by SEC. Typically, injecting about 0.5–1 mg of polymer into the SEC and collecting — Preparatory investigations

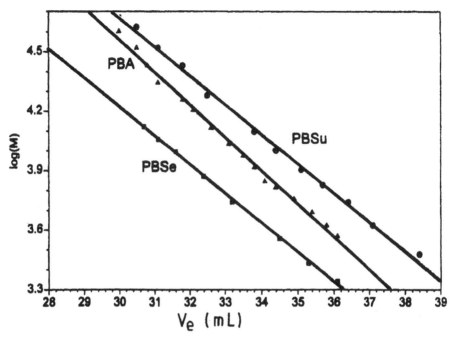

Fig. 7.6. Plots of log molar mass vs elution volume for PBSe (■), PBA (▲), and PBSu (●) (Reprinted from [32] with permission of American Chemical Society, U.S.A.)

MALDI-TOF analysis

25–100 fractions yields sufficient quantities for MALDI-TOF analysis. Selected nearly monodisperse fractions are analyzed by MALDI-TOF and the data are used for constructing the corresponding log M vs. elution volume curves. Figure 7.6 shows the calibration lines obtained for PBA, PBSu, and PBSe.

As can be seen clearly from these plots, the calibration lines of the different polyesters are different in position and in slope. Apparently, the hydrodynamic volumes lay in the order PBSe > PBA > PBSu, showing a correlation with their chemical structure. The molar masses calculated based on the SEC-MALDI-TOF procedure are given in Table 7.2.

Copolyesters which were prepared from 50:50 mixtures of different dimethyl esters are fractionated in a similar way. Figure 7.7 shows an illustrative example of the SEC trace of polybutylene adipate/sebacate (PBA-Se) together with the MALDI-TOF spectra of fractions taken at different elution volumes. The corresponding calibration curve for the copolymer together with the calibration lines for the homopolymers is given in Fig. 7.8A. It is clear from this plot that the copolymer deviates over its entire molar mass range from the additivity of hydrodynamic volumes, which is normally assumed in conventional SEC experiments. The same behavior is obtained for the copolymer polybutylene adipate/succi-

Table 7.2. Molar mass distributions of polyesters determined by SEC-MALDI-TOF

Sample	Feed[a]	M_w (SEC)[b]	M_w^c	M_n^c	M_w/M_n
PBA			14,000	8,000	1.75
PBSe			17,000	10,000	1.70
PBSu			15,400	9,200	1.67
PBA-Se	50:50	13,200	12,800	8,500	1.51
PBA-Su	50:50	14,200	11,600	6,900	1.68
PBSe-Su	50:50	15,400	8,600	5,400	1.59
PBA-Se-Su	33:33:33	8,400	15,000	8,300	1.81

[a] Molar ratio of ester monomers
[b] Determined by SEC with universal calibration
[c] Determined by SEC-MALDI-TOF

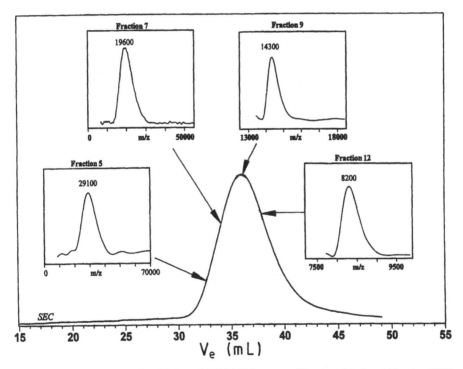

Fig. 7.7. SEC chromatogram for PBA-Se and MALDI-TOF spectra of fractions 5, 7, 9, and 12, solvent THF. (Reprinted from [32] with permission of American Chemical Society, U.S.A.)

nate (PBA-Su). In this case, the calibration line for the copolymer is even lower than the calibration lines of the corresponding homopolymers; see Fig. 7.8B.

Comparing the molar mass data obtained by conventional SEC with the data from SEC-MALDI-TOF, significant deviations are

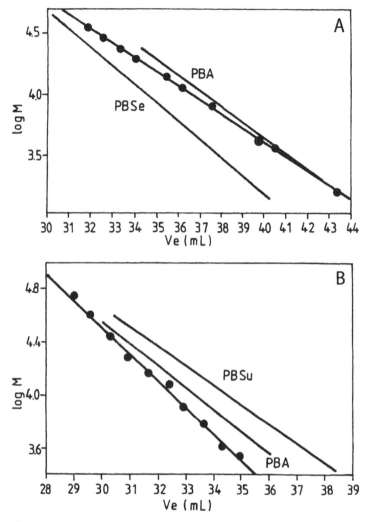

Fig. 7.8A,B. Plots of log molar mass vs elution volume for: **A** PBSe, PBA, and PBA-Se. **B** PBA, PBSu, and PBA-Su. (Reprinted from [32] with permission of American Chemical Society, U.S.A.)

obtained for PBSe-Su and PBA-Se-Su. This is a further strong indication that the hydrodynamic properties of the copolymers are more complex than can be extrapolated from the behavior of the homopolymers.

7.3.3 Poly(ethylene oxides) [28]

Aim

One of the most important classes of functional homopolymers is the class of alkyloxy and aryloxy poly(alkylene oxides). These

7.3 Off-line Measurement of LC Fractions

oligomers and polymers are in widespread use as surfactants. Depending on their molar mass and the chemical structure of the terminal groups the amphiphilic properties change, thus influencing the surface activity. Due to the different initiation, chain transfer and chain termination mechanisms, and possible impurities in the reaction mixture, species having different terminal groups bound to the PEO chain are formed. Very frequently, the surfactant properties are tuned by using mixtures of different alcohols or phenols as the starting materials. To elucidate the structure-property relationship of the products, it is important to know the chemical structure and the number of terminal groups in addition to the molar mass or oligomer distribution. This dual information can be obtained by separating chromatographically according to the endgroups and then analyzing the oligomer distributions by MALDI-TOF MS.

$$Alk(Ar)\text{-}OH + \underset{O}{CH_2\text{-}CH_2} \longrightarrow Alk(Ar)\text{-}(OCH_2CH_2)_n\text{-}OH$$

Secondary reactions ⇨ ⇨ ⇨ additional functionality fractions

$$H\text{-}(OCH_2CH_2)_n\text{-}OH$$

$$\overline{\text{-}(OCH_2CH_2)_n\text{-}}$$

$$Alk(Ar)\text{-}(OCH_2CH_2)_n\text{-}OAlk(Ar)$$

Technical polyethylene oxides with fatty alcohol endgroups (FAE). **Materials** samples 1 (C_{10}-PEO), 2 (C_{12}-PEO), 3 (C_{13}-PEO), 4 ($C_{13,15}$-PEO)

Chromatographic system — modular HPLC system compris- **Equipment** ing a Waters model 510 pump, a Rheodyne six-port injection valve, and a Waters column oven.

Columns — Nucleosil RP-18 of Macherey-Nagel, 5 µm average particle size and 100 Å average pore diameter. Column size was 125×4 mm I.D.

Mobile Phase — mixture of acetonitrile and water 70:30 v/v, all solvents are HPLC grade.

Detector — Waters differential refractometer R 410.

Sample amount — 20 to 50 µl of a 4 mg/ml solution in the mobile phase.

MALDI-TOF system — Kratos Kompact MALDI 3 with linear and reflectron detectors, acceleration voltage of positive 20 kV.

Preparatory investigations

MALDI-TOF sample preparation — the samples were directly taken from the chromatographic separations and mixed with the matrix 2,5-dihydroxy benzoic acid in water. Then 0.5 µl of the mixture was droped on the MALDI sample target, dried and analyzed.

Liquid chromatography can be used to separate heterogeneous polymers according to molar mass or chemical composition. Liquid chromatography at the critical point of adsorption (LC-CC) is frequently used to separate functional homopolymers with respect to their endgroups. In this case, the chromatography occurs independently of molar mass solely according to functionality and, as a result, a functionality type distribution is obtained [33]. Another option is the separation with respect to functionality *and* molar mass. Such separation can be obtained by using a stationary phase interacting preferentially with the endgroups and a mobile phase promoting weak adsorption. As has been shown previously, for alkyloxy/aryloxy-terminated polyethylene oxides a suitable chromatographic system is a reversed (RP) stationary phase combined with acetonitrile-water as the eluent.

The separation of a series of fatty alcohol ethoxylates (FAE) by endgroup and molar mass is presented in Fig. 7.9. As can be seen, for the C_{12}- and $C_{13,15}$-alkyl terminated FAE, very delicate separations are obtained. A first sharp and uniform peak is obtained at a retention time of 60 s followed by broader peaks that show a substructure. The first peak can be assigned to the by-product polyethylene glycol (PEG), while the broad peaks correspond to the FAE fractions. Obviously, the substructures are due to oligomer separations within each functionality fraction. Accordingly, each peak of the substructures corresponds to one single oligomer with a specific endgroup and degree of oligomerization.

MALDI-TOF analysis

For properly separated FAE samples different parameters may be obtained quantitatively from the chromatograms. First of all, the amount of the different functionality fractions may be determined from the relative peak areas, taking into account the different detector responses. For the determination of the molar mass distribution of each functionality fraction, preparative separations may be carried out and the resulting fractions may be investigated by SEC or any other method for molar mass determination.

For samples that are separated with respect to functionality and oligomer distribution, see Fig. 7.9, samples 2 (B) and 4 (D), a different approach can be used. From the analytical separation, fractions of the single oligomers within each functionality fraction can be collected and subjected to MALDI-TOF mass spectrometry for identification. The resulting mass spectrum yields

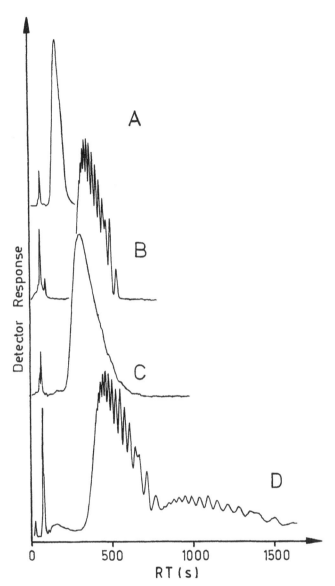

Fig. 7.9. Chromatograms of FAE samples 1–4 in a weak adsorption mode, stationary phase: Nucleosil RP-18, 125 × 4 mm I.D., mobile phase: acetonitrile-water 70:30 v/v, samples C_{10}- (A), C_{12}- (B), C_{13}- (C), $C_{13,15}$-PEO (D)

information on the type of endgroup and the degree of oligomerization.

In order to assign the oligomer peaks in the chromatograms of samples 2 and 4, chromatographic separations are conducted and the oligomer fractions are collected, resulting in amounts of 5–20 ng substance per fraction in the mobile phase. The solutions are taken directly, mixed with the matrix solution, and subjected to the MALDI- TOF MS experiments.

For sample 2 (C_{12}-PEO) 14 fractions are collected, fraction 1 being PEG and fractions 2–14 containing the C_{12}-terminated ethylene oxide oligomers. The resulting spectra of some of the fractions are shown in Fig. 7.10.

The MALDI MS spectrum of fraction 1 consists of two peak series, one representing the $[M+Na]^+$ molecular ions and the other representing the $[M+K]^+$ molecular ions of the PEG oligomers. The intensity of the peaks in this case is equivalent to concentration and from the relative concentrations of the oligomers the molar mass distribution of the PEG fraction can be calculated. The MALDI MS spectra of fractions 2–14 show one major peak each, representing the $[M+Na]^+$ molecular ion of the corresponding oligomer, and some minor peaks of neighbor oligomers due to incomplete chromatographic separation. In all cases the major peak and its corresponding mass is used to assign a degree of oligomerization to the corresponding peak in the chromatogram.

Similarly, the oligomer peaks in the $C_{13,15}$-PEO (sample 4) are collected and analyzed by MALDI-TOF; see Fig. 7.11. Fraction 1 again can be assigned to PEG, while the oligomer distributions (fractions 2 and 3) belong to fatty alcohol-terminated ethylene oxide oligomers. From the masses assigned to the peaks and the peak-to-peak mass increment of the ethylene oxide repeating unit the mass of the endgroup for the different fractions may be calculated:

Fraction 1
$$[M+Na]^+ = 41 + 44n$$
$$M = 18 + 44n$$
$$H(EO)_nOH$$

Fraction 2
$$[M+Na]^+ = 223 + 44n$$
$$M = 200 + 44n$$
$$C_{13}H_{27}(EO)_nOH$$

Fraction 3
$$[M+Na]^+ = 251 + 44n$$
$$M = 228 + 44n$$
$$C_{15}H_{31}(EO)_nOH$$

The endgroups of fractions 1–3 can be identified as being polyethylene glycol (PEG) (α,ω-dihydroxy endgroups), C_{13}-terminated polyethylene oxide (PEO) (α-tridecyl-ω-hydroxy endgroups), and C_{15}-terminated PEO (α-pentadecyl-ω-hydroxy endgroups), re-

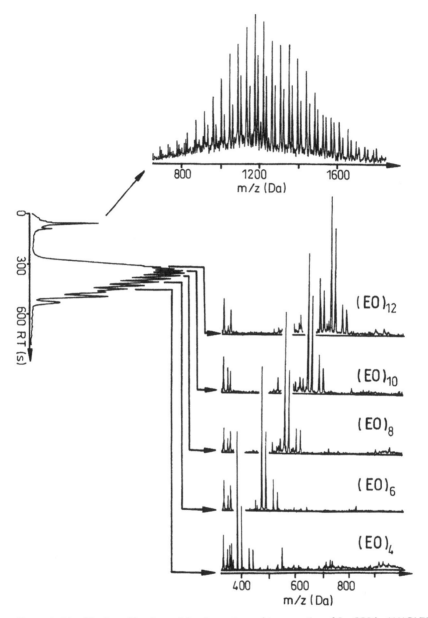

Fig. 7.10. Identification of fractions of the chromatographic separation of C_{12}-PEO by MALDI-TOF MS, chromatogram taken from Fig. 7.9. (Reprinted from [33] with permission of Springer, Berlin Heidelberg New York)

spectively. Using MALDI MS, the oligomer distribution of the PEG fraction is measured directly. The MALDI MS spectrum of the PEG fraction (fraction 1) is given in Fig. 7.12. The homologous series of higher peak intensity corresponds to the $[M+K]^+$ molecular ions of the ethylene glycol oligomers. For fractions 2 and 3

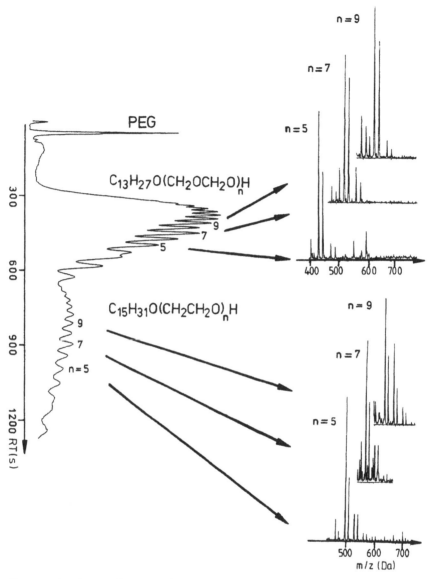

Fig. 7.11. Identification of fractions of the chromatographic separation of $C_{13,15}$-PEO by MALDI-TOF MS, chromatogram taken from Fig. 7.9; peak assignment indicates degree of oligomerization (n). (Reprinted from [28] with permission of Elsevier Science B.V., The Netherlands)

more or less homogeneous oligomer fractions are obtained which reveal information on the type of the endgroup and the degree of oligomerization; see Fig. 7.11.

From the assignment of the oligomer peaks in the chromatograms, oligomer calibration curves molar mass vs retention time are obtained for the C_{12}-, C_{13},- and C_{15}-fractions; see Fig. 7.13.

7.3 Off-line Measurement of LC Fractions

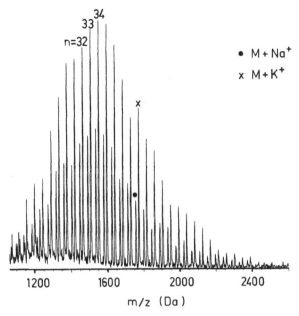

Fig. 7.12. MALDI-TOF spectrum of fraction 1 (PEG) from chromatographic separation shown in Fig. 7.11; peak assignment indicates degree of oligomerization (n). (Reprinted from [28] with permission of Elsevier Science B.V., The Netherlands)

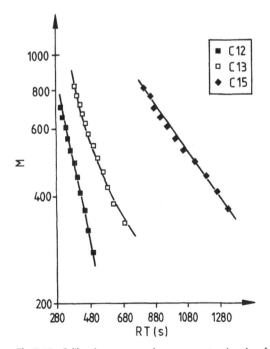

Fig. 7.13. Calibration curves molar mass vs retention time for the functionality fractions of samples 2 and 4, corresponding chromatograms see Figs. 7.10 and 7.11. (Reprinted from [33] with permission of Springer, Berlin Heidelberg New York)

With these calibration curves, the average molar masses and the polydispersities of the functionality fractions can be calculated.

7.3.4 Block Copolymers of Ethylene Oxide and Propylene Oxide

Aim Random and block copolymers based on ethylene oxide (EO) and propylene oxide (PO) are important precursors for polyurethanes. Their detailed chemical structure, i.e., the chemical composition, block length, and molar mass of the individual blocks, may determine the properties of the final product. For triblock copolymers $HO(EO)_n(PO)_m(EO)_nOH$ the detailed analysis relates to the determination of the total molar mass and the degrees of polymerization of the inner PPO block (m) and the outer PEO blocks (n).

Materials Technical triblock copolymer with an average structure of $HO(EO)_n(PO)_m(EO)_nOH$. The sample under investigation has been prepared by anionic polymerization at 110 °C using potassium glycolate as initiator.

Equipment Chromatographic System — modular HPLC system comprising a Waters model 501 pump, a Rheodyne six-port injection valve and a Waters column oven.

Columns — Nucleosil RP-18 of Macherey-Nagel, 5 μm average particle size and 100Å average pore diameter. Column size was 250×4 mm I.D.

Mobile phase — mixture of acetonitrile and water 42:58 v/v, all solvents are HPLC grade.

Detector — Waters differential refractometer R 410.

Sample Amount — 20 to 50 μl of a 5–10 mg/ml solution in the mobile phase.

MALDI-TOF system — Kratos Kompact MALDI 3 with linear and reflectron detectors, acceleration voltage of positive 20 kV.

MALDI-TOF sample preparation — the samples were directly taken from the chromatographic separations and mixed with the matrix 2,5-dihydroxy benzoic acid in water. Then 0.5 μl of the mixture was deposited on the MALDI sample target, dried and analyzed.

Preparatory investigations It was shown in previous investigations that block copolymers of ethylene oxide and propylene oxide may be separated with respect to the chain length of the propylene oxide block by liquid chromatography at the critical point of adsorption [34,35]. Operating at chromatographic conditions corresponding to the critical point of PEG, the ethylene oxide block behaves chromatographically as "invisible" and retention of the block copolymer is solely directed by the propylene oxide block. Critical conditions for PEO

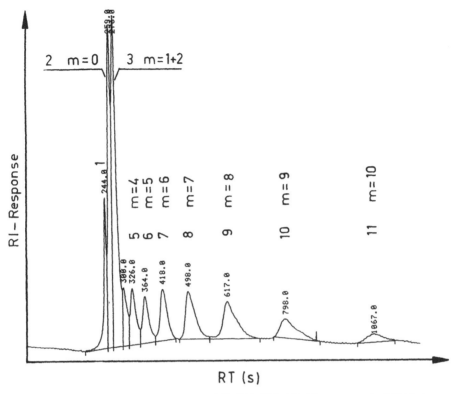

Fig. 7.14. Separation of a triblock copolymer H(EO)$_n$(PO)$_m$(EO)$_n$OH with respect to the PPO block; peak assignment indicates fraction and degree of polymerization of the PPO block (m); stationary phase: Nucleosil RP-18; mobile phase: acetonitrile-water 42:58 v/v (Reprinted from [33] with permission of Springer, Berlin Heidelberg New York)

can be established on an RP-18 stationary phase when a mobile phase of acetonitrile-water 42:58 v/v is used. Figure 7.14 represents the separation of the triblock copolymer H(EO)$_n$(PO)$_m$(EO)$_n$OH at the critical point of the ethylene oxide blocks.

The assignment of the peaks is based on comparison with the chromatogram of a polypropylene glycol. The first and the second peak in the chromatogram, at retention times of 244 s and 259 s respectively, could not be identified directly, whereas the third peak at a retention time of 275 s corresponds to m = 1–2, m being the degree of polymerization with respect to propylene oxide. The peak at 300 s corresponds to m = 3, the peak at 326 s corresponds to m = 4, and so on. Accordingly, every peak is uniform with respect to m but has a distribution in block length with respect to the polyethylene oxide blocks (n).

The liquid chromatographic separation of the block copolymer yields information on the degree of oligomerization (m) with regard to PPO. It does not yield information on the chain lengths of

MALDI-TOF analysis

the PEO blocks. In addition, the first two peaks in the chromatogram need further analysis, since they could not be assigned by comparison with PPG.

Therefore, in order to identify peaks 1 and 2 in Fig. 7.14 and to determine n of the other fractions, they are collected and subjected to MALDI MS. For these experiments a small amount of lithium chloride is added to the sample, thus favoring the formation of $[M+Li]^+$ molecular ions. The MALDI MS spectra of fractions 1-3 are summarized in Fig. 7.15.

For fraction 1 a number of peaks in the mass range up to 500 Da is obtained, the peaks at 284 Da and 485 Da being the most intense ones. However, a homologous series, representing ethylene oxide oligomers, is not obtained. Therefore, these peaks are assumed to appear due to impurities in the reaction product or the MALDI MS matrix. Much more informative is the spectrum of fraction 2, showing the expected series of peaks having a peak-to-peak mass increment of 44 Da. This increment exactly equals the mass of the ethylene oxide repeating unit and, accordingly, the observed series of peaks represents an ethylene oxide oligomer series. From the masses assigned to the peaks and the peak-to-peak mass increment of the ethylene oxide unit, the mass of the endgroup is calculated to be 18 Da:

Fraction 2 $\quad [M+Li]^+ = 25 + 44n$

$M = 18 + 44n$

Therefore, fraction 2 represents polyethylene glycol (m=0). Due to poor chromatographic separation, in addition to the PEG peaks the major peaks of fraction 1 at 284 Da and 485 Da are detected as well.

The spectrum of fraction 3 is very complex. In addition to some minor PEG peaks, two other homologous series with a peak-to-peak increment of 44 Da are obtained. These may be assigned to the following structures:

(●) $\quad [M+Li]^+ = 83 + 44n$

$M = 76 + 44n$

$H(EO)_n(PO)_1(EO)_nOH$

(X) $\quad [M+Li]^+ = 141 + 44n$

$M = 134 + 44n$

$H(EO)_n(PO)_2(EO)_nOH$

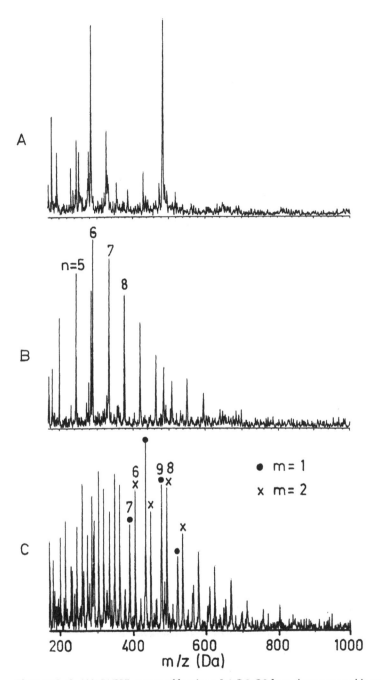

Fig. 7.15 A-C. MALDI-TOF spectra of fractions: **A** 1; **B** 2; **C** 3 from chromatographic separation shown in Fig. 7.14; peak assignment indicates degree of oligomerization of PEO (n). (Reprinted from [28] with permission of Elsevier Science B.V., The Netherlands)

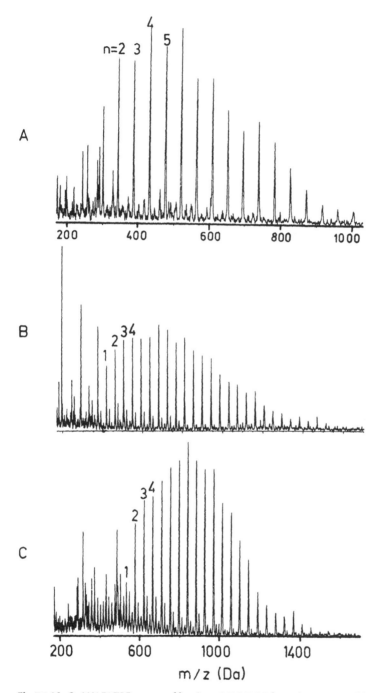

Fig. 7.16 A–C. MALDI-TOF spectra of fractions: **A** 5; **B** 7; **C** 9 from chromatographic separation shown in Fig. 7.14; peak assignment indicates degree of oligomerization of PEO (n). (Reprinted from [28] with permission of Elsevier Science B.V., The Netherlands)

7.3 Off-line Measurement of LC Fractions

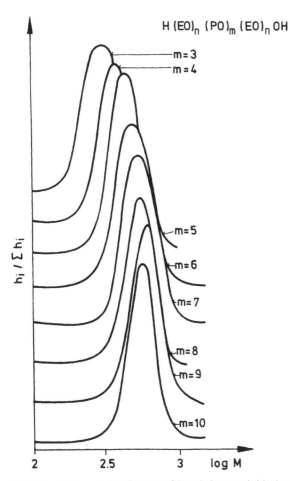

Fig. 7.17. Molar mass distributions of the ethylene oxide blocks of fractions 4–11 as determined by MALDI-TOF. (Reprinted from [28] with permission of Elsevier Science B.V., The Netherlands)

Table 7.3. Assignment of fractions 4–11 from chromatographic separation of the block copolymer

Fraction	m	Structure
4	3	$H(EO)_n(PO)_3(EO)_nOH$
5	4	$H(EO)_n(PO)_4(EO)_nOH$
6	5	$H(EO)_n(PO)_5(EO)_nOH$
7	6	$H(EO)_n(PO)_6(EO)_nOH$
8	7	$H(EO)_n(PO)_7(EO)_nOH$
9	8	$H(EO)_n(PO)_8(EO)_nOH$
10	9	$H(EO)_n(PO)_9(EO)_nOH$
11	10	$H(EO)_n(PO)_{10}(EO)_nOH$

Accordingly, fraction 3 represents a mixture of the homologous series with m = 1 and m = 2.

Due to better separation of the following fractions, their MALDI MS spectra each represent one homologous series; see Fig. 7.16 for fractions 5, 7, and 9.

The assignment of fractions 4–11 taken from the chromatographic separation is given in Table 7.3.

From the intensities of the individual peaks in the MALDI MS spectra the relative abundance of each oligomer can be determined. Thus, for each fraction the oligomer or molar mass distribution with respect to the ethylene oxide blocks can be calculated; see Fig. 7.17.

7.3.5 Epoxy Resins [37]

Aim

Epoxy resins represent one of the most important types of polymers for the production of three-dimensional network materials. They are mostly prepared by the reaction of epichlorohydrin and bisphenol A (BPA). As a result of this reaction oligomers are formed which mainly contain glycidyl endgroups. The functionality with respect to the glycidyl groups can approach values greater than 1.9, but this is largely a function of controlling side reactions, such as hydrolysis of epoxy groups, incomplete dehydrohalogenation, and abnormal addition of epichlorohydrin. Epoxy resins with varying degrees of polymerization (k) are prepared by two processes. In the Taffy process epichlorohydrin, bisphenol A and a stoichiometric amount of sodium hydroxide are reacted to yield oligomers with odd and even degrees of polymerization; see Scheme 7.1.

An alternative method is the chain-extension reaction of crude bisphenol A diglycidyl ether with bisphenol A, frequently referred to as the advancement process. In the advancement process, branching occurs by reaction of the hydroxy groups in the polymer backbone with the epoxy functional groups.

The analysis of epoxy resins is complicated by the fact that, in addition to molar mass distribution (MMD), a functionality type distribution (FTD) and a topological distribution (linear vs branched molecules) are encountered. The structural heterogeneity of epoxy resins including FTD can be studied by LC-CC [36, 37]. Under appropriate chromatographic conditions a separation according to the endgroups irrespective of the molar mass distribution can be conducted. The analysis of the functionality fractions with regard to oligomer distribution shall be performed by MALDI-TOF mass spectrometry.

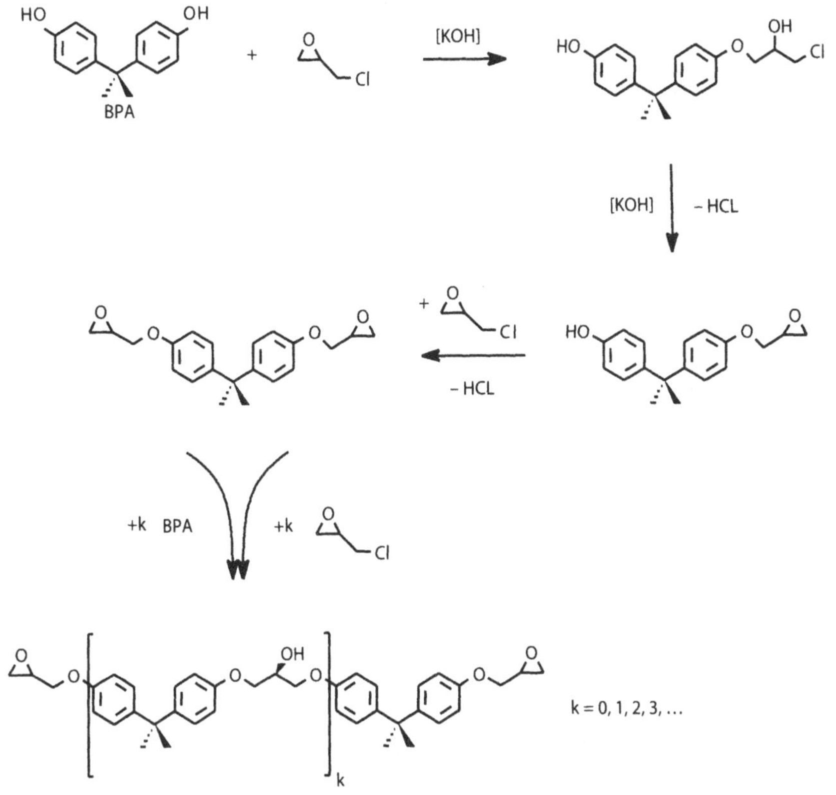

Scheme 7.1.

The epoxy resins under investigation are research products of Dow Deutschland Inc., Rheinmünster, Germany. Their average molar masses determined by SEC are summarized in Table 7.4.

Chromatographic system — modular HPLC system comprising a TSP model 100 HPLC pump, a Rheodyne six-port injection valve, and a Waters column oven.

Columns — Nucleosil 50-5 of Macherey-Nagel, 5 μm average particle size. Column size was 200×4 mm I.D.

Mobile Phase — mixture of THF and *n*-hexane 74:26 v/v, all solvents are HPLC grade.

Detector — Waters tunable UV detector model 486, wavelength 280 nm.

Sample amount — 10–20 μl of a 5 mg/ml solution in the mobile phase.

MALDI-TOF system — Kratos Kompact MALDI 3 with linear and reflectron detectors, acceleration voltage of positive 20 kV.

MALDI-TOF sample preparation — the samples were taken from the chromatographic separations, the solvent evaporated and the residue mixed with the matrix 2,5-dihydroxy benzoic acid

Materials

Equipment

Table 7.4. Average molar masses and polydispersities of epoxy resins, determined by SEC

Sample	M_n [g/mol]	M_w [g/mol]	Polydispersity $Q\,(M_w/M_n)$
1	5020	23,800	4.74
2	3280	11,410	3.48
3	2890	9,560	3.31
4	4090	16,190	3.96
5	2650	6,890	2.60
6	3920	16,150	4.12
7	1060	2,330	2.20
8	860	1,510	1.76

in THF. Then 0.5 μl of the mixture was dropped on to the MALDI sample target, dried, and analyzed.

Preparatory investigations

Epoxy resins are complex oligomer mixtures which are distributed with respect to molar mass, endgroup functionality, and branching. Using different techniques of liquid chromatography, quantitative information on these parameters can be obtained. For a first information on molar mass, the samples are analyzed by SEC. The SEC chromatograms of different epoxy resins are presented in Fig. 7.18. While for sample 8 well resolved oligomer

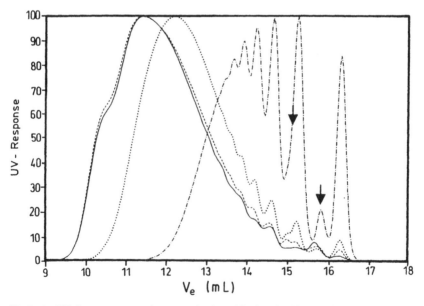

Fig. 7.18. SEC chromatograms of epoxy resins, (——) Resin 4, (·····) Resin 5, (– – –) Resin 6, (_._.) Resin 8; stationary phase: PL Mixed-D + Mixed-E; mobile phase: THF. (Reprinted from [37] with permission of Wiley-VCH, Germany)

peaks are obtained, indicating rather low molar masses, the other samples show more or less continuous elution profiles due to their significantly higher molar masses. The last peak in the chromatograms at 16.3 ml corresponds to the monomeric bisglycidyl bisphenol A, while the peaks at 15.3 ml, 14.8 ml, and 14.2 ml can be assigned to the respective dimer, trimer, and tetramer. Since no calibration standards for epoxy resins are available, a calibration curve for calculating the MMDs must be constructed from polystyrene calibration standards and epoxy resin oligomers.

The individual epoxy oligomers are obtained by preparatively separating sample 8. The molar masses of the oligomers are then determined by MALDI-TOF mass spectrometry. Figure 7.19 shows the polystyrene and the epoxy oligomer calibration curves and indicates that at molar masses less than 2000 g/mol there is a significant difference between the curves. However, above molar masses of 3000 g/mol the calibration curves merge into one common calibration curve. The reason for the differences in the low molar mass range is the significantly different stiffness of oligostyrene and epoxy oligomers. The higher stiffness of the epoxy oligomers results in higher hydrodynamic volumes and lower elution volumes, accordingly. For the calculation of the MMDs a combination of both calibration curves is used. The lower molar mass range was calibrated with the epoxy oligomers, while above a molar mass of 3000 g/mol the polystyrene calibration curve was

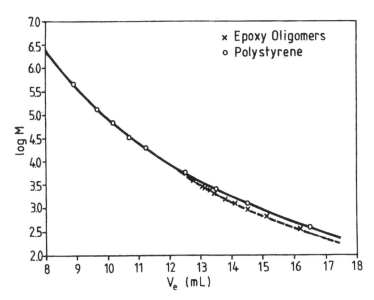

Fig. 7.19. SEC calibration curves using polystyrene (**X**) and epoxy oligomers (O) (Reprinted from [37] with permission of Wiley-VCH, Germany)

used. The average molar masses and polydispersities of the samples are summarized in Table 7.4.

Another interesting feature of the chromatograms was that, in addition to the major oligomer peaks, small peaks were obtained which did not fit into the oligomer series; see arrows in Fig. 7.18. These peaks indicate the presence of oligomers with chemical structures, other than the major peaks, presumably endgroups. SEC is not sufficiently selective to separate these functionally defective molecules. A very efficient method for such separations is LC-CC.

In general, the LC-CC conditions are determined by running a number of calibration samples of different molar masses in eluents of different compositions. This was not possible with the epoxy resins, since calibration samples were not available. However, it is known that the major functionality fraction in epoxy resins is always the bisglycidyl fraction, constituting the major elution peak in the chromatograms. This peak was regarded as the reference peak and its chromatographic behavior was investigated as a function of the eluent composition. As can be seen in Fig. 7.20, significantly different chromatograms were obtained depending on the composition of the eluent. In pure THF, the sample exhibited typical SEC separation behavior, as is indicated by the oligomer peaks. With increasing concentrations of n-hexane in the mobile phase, adsorptive interactions were promoted and

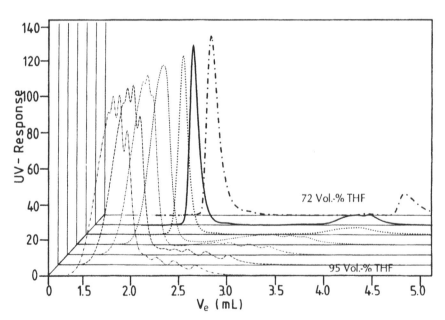

Fig. 7.20. Chromatograms of sample 8 at different eluent compositions; stationary phase: Nucleosil 50-5; mobile phase: THF-n-hexane, 95/90/85/80/75/74/72 % THF. (Reprinted from [37] with permission of Wiley-VCH, Germany)

the elution peak maximum shifted to higher elution volumes. In addition, the oligomer peaks merged in one narrow elution peak and a part of the sample eluted as a separate peak at high elution volumes. The critical point of adsorption, corresponding to a molar mass independence of the elution volume, was obtained at an eluent composition of THF-n-hexane 74:26 v/v. At these conditions, separation is solely directed by functionality and branching.

A typical chromatogram at critical conditions is given in Fig. 7.21. As can be seen, three well separated elution regions P1, P2, and P3 were obtained. P1–P3 correspond to fractions of different chemical composition, and from the chromatographic conditions it can be assumed that P1 contains the least polar constituents, while P3 contains the most polar constituents. Since in most of the samples P1 constituted the highest peak, it was preliminarily assigned to the bisglycidyl oligomer fraction. Due to hydrolysis of the epoxy groups, diol endgroups can be present in the samples in addition to the epoxy groups. For a detailed analysis, sample 8 is fractionated preparatively into fractions A–H, which are analyzed by MALDI-TOF mass spectrometry.

For the analysis of fractions A–H obtained by preparative separation of sample 8, 2,5-dihydroxy benzoic acid was used as the matrix. As is known from previous experiments, the epoxy oligomers are ionized by the attachment of Na^+-cations resulting in

MALDI-TOF analysis

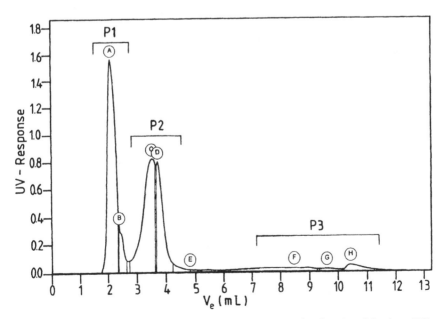

Fig. 7.21. LC-CC chromatogram of sample 8; stationary phase: Nucleosil 50-5; mobile phase: THF-n-hexane 74:26 v/v, detection: UV 280 nm. (Reprinted from [37] with permission of Wiley-VCH, Germany)

the formation of [M+Na]$^+$ molecular ions. Epoxy resin based oligomer series can be identified by typical peak-to-peak mass increments of 284 Da, being the mass of the repeat unit.

The MALDI-TOF mass spectra of fractions A, C, and D are presented in Fig. 7.22. All spectra exhibit epoxy resin based oligomer series, as expected. For fraction A one major peak series is obtained (**X**) which can be assigned to the bisglycidyl oligomers. The maximum of the oligomer series is centered at about 900 Da, being in good agreement with the M_n value of 860 g/mol determined by SEC for sample 8. In addition to this peak series a second series of low intensity is observed (□), exhibiting a mass difference of 57 Da towards the major peak series. This mass difference is typically for branched epoxy resin chains, where one branching point per chain is located at a secondary hydroxy group.

Therefore this oligomer series can be assigned to singly branched fully epoxydized macromolecules. This interpretation agrees well with the shift towards higher molar masses observed for series □.

While fraction B is very similar to A, for fraction C a MALDI-TOF spectrum with only one oligmer series is obtained (Δ). Due to incomplete separation this oligomer series is also observed in fraction D. The major oligomer series in fraction D (●) can be assigned to linear epoxy resin molecules, where one glycidyl group was hydrolyzed to a diol group. Similar to fraction A, the oligomer series Δ exhibits a mass difference of 57 Da towards peak series ●. Accordingly, series Δ can be assigned to singly branched α-glycidyl-ω-diol oligomers, having a glycidyl endgroup at the branch.

The MALDI-TOF spectra of fractions F, G, and H are shown in Fig. 7.23. Similar to the interpretation of the spectra for fractions A–D, linear and branched oligomer molecules can be identified from the typical mass increments of 57 Da. Peak series ■ corresponded to linear bisdiol terminated oligomers, while series σ and ∞ can be assigned to singly and doubly branched oligomers, respectively. It is obvious from the mass spectra that separation in LC-CC is in the order of decreasing branch numbers, i.e., first the

7.3 Off-line Measurement of LC Fractions

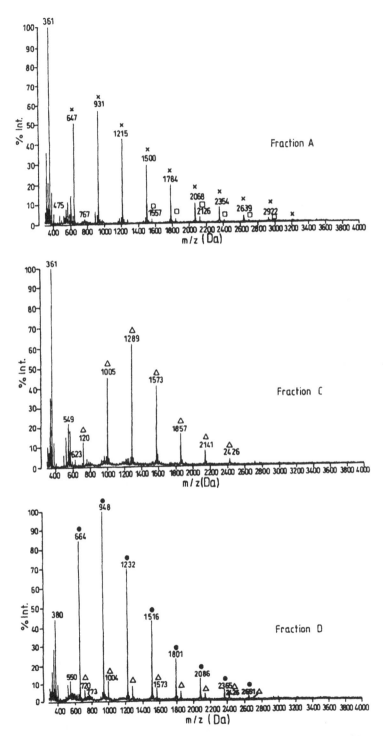

Fig. 7.22. MALDI-TOF spectra of fractions A, C, and D from the chromatographic separation of sample 8. (Reprinted from [37] with permission of Wiley-VCH, Germany)

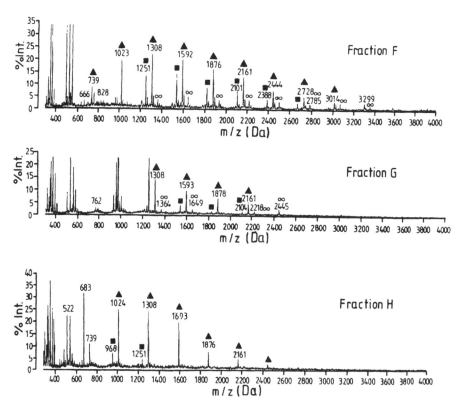

Fig. 7.23. MALDI-TOF spectra of fractions F, G, and H from the chromatographic separation of sample 8. (Reprinted from [37] with permission of Wiley-VCH, Germany)

doubly branched oligomers are eluted, followed by the singly branched and linear oligomers. The complete assignment of the oligomer series in the chromatographic fractions of sample 8 is given in Table 7.5. From this assignment it can be deducted that LC-CC elution range P1 contains all fully epoxydized oligomers, while P2 and P3 are due to monodiol and bisdiol oligomers, respectively.

To summarize, using different liquid chromatographic techniques and MALDI-TOF mass spectrometry, commercial bisphenol A based epoxy resins can be analyzed with respect to molar mass, functionality and branching. By LC-CC it is possible to separate the samples into functionally, homogeneous fractions. The chemical composition of the fractions can be determined by MALDI-TOF MS giving additional information on the presence of singly and doubly branched molecules. SEC is a useful technique for determining the total molar mass of the samples, provided a proper calibration curve is established. However, using LC-CC, MALDI-TOF, and SEC as separate methods, corresponding infor-

7.3 Off-line Measurement of LC Fractions

Table 7.5. Assignment of epoxy resin structures to the chromatographic fractions of sample 8

Fraction	General structure	Code	Mass peak $[M + Na]^+$
A, (B)		□	$704+(284.37 \cdot k)$
		X	$647+(284.37 \cdot k)$
C		Δ	$723+(284.37 \cdot k)$
D		●	$666+(284.37 \cdot k)$
F;G		∞	$797+(284.37 \cdot k)$
		σ	$740+(284.37 \cdot k)$
H		■	$683+(284.37 \cdot k)$

mation on MMD, FTD, and branching is not easily accessible. For obtaining corresponding MMD and FTD data it is more feasible to combine LC-CC and SEC or MALDI-TOF in an on-line set-up.

7.3.6 Block Copolymers of L-Lactide and Ethylene Oxide [38]

Poly(ethylene oxide)-*b*-poly(L-lactide)s (PEO-*b*-PLA) are water-soluble block copolymers with a hydrophilic PEO block coupled to a hydrophobic PLA block. Such block copolymers are of increasing interest in the field of pharmaceutical and biomedical ap-

Aim

plications because of their biodegradability and biocompatibility. Various physical properties, e.g., micellization, need to be considered when optimizing the functional characteristics of the block copolymers. In particular, the balance of hydrophilic and hydrophobic components in the block copolymers is crucial. Consequently, the accurate molecular characterization including the determination of the total molar mass, the molar masses of the blocks, and the chemical composition is highly relevant.

Interaction chromatography is the most prominent method for the determination of the heterogeneity of complex polymers. In particular, LC-CC is a versatile technique for analyzing block copolymers with regard to the separate blocks; see also Sect. 7.3.4. Operating at critical conditions of one block, this block is made chromatographically "invisible" while the other block can be analyzed with respect to chain length distribution [33,39-42].

In the present application, PEO-b-PLA shall be separated under chromatographic conditions corresponding to the critical point of PEO. The resulting fractions shall then be analyzed by MALDI-TOF MS to yield structural information on the molar masses of the PEO blocks.

Materials

The PEO-PLA block copolymer under investigation is a research product of POSTECH, Pohang, Korea. The PEO block has a molar mass of $M_n = 2000$ g/mol determined by SEC, the total molar mass was calculated from ^{1}H-NMR to be roughly 2700 g/mol.

Equipment

Chromatographic System — modular HPLC system comprising an LDC constaMetric 3200 HPLC pump, and a Rheodyne six-port injection valve. Column temperature was controlled by a thermostated column jacket.

Columns — Phenomenex Luna C18, 100 Å average pore size. Column size was 250×4.6 mm I.D.

Mobile phase — mixture of acetonitrile (ACN) and water 60:40 v/v, all solvents are HPLC grade.

Detector — Shodex RI 71 refractometer and LDC spectroMonitor 3200 UV detector.

Sample amount — 10–20 µl of a 5 mg/ml solution in the mobile phase

MALDI-TOF system — PerSeptive Biosystem Voyager Elite MALDI-TOF MS, acceleration voltage of positive 25 kV.

MALDI-TOF sample preparation — after HPLC fractionation the samples were dried and then diluted with THF. The solutions were mixed with the matrix 2,5-dihydroxy benzoic acid in THF. Then 1 µl of the mixture was pipetted on to the MALDI sample target, dried, and analyzed.

Preparatory Investigations

The critical conditions for PEO can be determined experimentally (a) by adjusting the eluent composition, as has been discussed in Sect. 7.3.4, or (b) by adjusting the temperature. Figure 7.24 displays the chromatograms of PEO-*b*-PLA at different temperatures (**A**) and different eluent compositions (**B**). The middle chromatograms display the separation at critical conditions of polyethylene glycol, corresponding to a column temperature of 68 °C and a mobile phase composition of ACN-water 60:40 v/v. At these conditions the block copolymer is separated solely with respect to the chain length distribution of the PLA block. In the present chromatographic regime PLA is eluted in an interac-

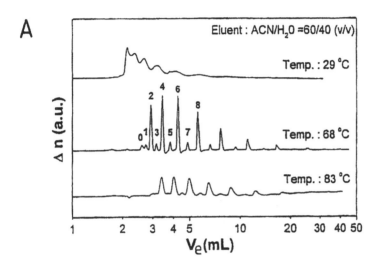

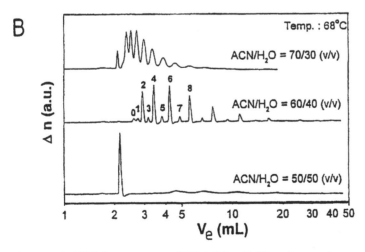

Fig. 7.24 A, B. HPLC chromatograms of PEO-*b*-PLA at: **A** different temperatures; **B** eluent compositions, stationary phase: Phenomenex Luna C18, eluent: ACN-water. (Reprinted from [38] with permission of American Chemical Society, U.S.A.)

tive mode of HPLC; accordingly longer PLA chain lengths correspond to higher retention volumes.

The labelling at the elution peaks corresponds to the degree of polymerization of the PLA block. Accordingly, peak 0 does not contain PLA units and corresponds to a fraction of PEO precursor block. Peak 1 corresponds to a degree of polymerization of $n=1$ for the PLA block, peak 2 corresponds to $n=2$, etc. Each peak in the chromatogram is monodisperse with regard to the PLA block but polydisperse with regard to the PEO block. The high intensity of the even-numbered elution peaks is due to the fact that the cyclic lactide dimer was used for the formation of the PLA block.

MALDI-TOF analysis

The analysis of the HPLC fractions by MALDI-TOF mass spectrometry yields information on the oligomer distribution of the PEO block. Since the fractions are monodisperse with respect to the chain length of the PLA block, the mass peak distributions presented in Fig. 7.25 must be solely due to the polydispersity of the PEO block.

This assumption is in perfect agreement with the major peak series found in the spectra. The peak-to-peak mass increment is 44 Da, corresponding to the mass of the ethylene oxide repeat unit. Ionization takes place via attachment of sodium cations and, therefore, the mass peaks correspond to the $[M+Na]^+$ molecular ions of the oligomers. Knowing that the general formula of the block copolymer is $CH_3O\text{-}(EO)_m\text{-}(LA)_n\text{-}H$, each oligomer peak can be assigned to a certain m and n by using the equation

$$[M+Na]^+ = 23 \,(Na) + 32 \,(endgroups) + 44.05 \,(EO) \times m + 72.06 \,(LA) \times n$$

Taking the mass peak at 2208 Da in Fig. 7.25A it can be calculated that the corresponding oligomer is composed of three LA groups ($n=3$) and 44 EO groups ($m=44$). The mass peak of 2281 Da in Fig. 7.25B corresponds to an oligomer with the same number of EO units but one more LA group ($n=4$).

A closer look at the spectra shows that the mass peaks appear as triplets of a large peak and two small peaks with mass increments of +16 Da and +32 Da. These minor peaks result from partial degradation of the PLA block, which is known to be easily degradable.

From the MALDI-TOF spectra the molar masses of the HPLC fractions can be calculated; see Table 7.6. Since the degree of polymerization for the PLA block is known for all these fractions, from the total molar masses the molar masses of the PEO blocks can be obtained. As can be expected, the molar masses of PEO blocks with different numbers of PLA residues are identical within experimental precision and are in good agreement with the SEC analysis of the precursor PEO.

7.3 Off-line Measurement of LC Fractions

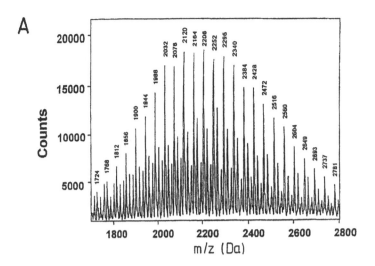

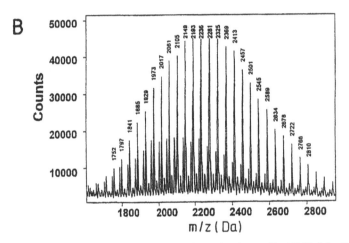

Fig. 7.25 A,B. MALDI-TOF spectra of PEO-b-PLA fractionated by LC-CC: **A** fraction 3; **B** fraction 4. (Reprinted from [38] with permission of American Chemical Society, U.S.A.)

Table 7.6. Molar mass characterization of PEO-b-PLA block copolymers by MALDI-TOF MS

Fraction	M_n (PEO-b-PLA) (g/mol)	M_n (PEO block) (g/mol)	M_w/M_n
1	2072	1977	1.026
2	2195	2028	1.027
3	2202	1963	1.026
4	2264	1953	1.027
5	2395	2012	1.019
6	2472	2017	1.023
7	2574	2047	1.021
8	2669	2070	1.014

7.3.7 Further Applications

Other examples of successful combinations of liquid chromatography and MALDI-TOF were given by Krüger et al., separating linear and cyclic fractions of polylactides by LC-CC [43]. Just and Krüger were able to separate cyclic siloxanes from linear silanols and to characterize their chemical composition [44]. The calibration of a SEC system by MALDI-TOF was discussed by Montaudo [45]. Polydimethyl siloxane (PDMS) was fractionated by SEC into different molar mass fractions. These fractions were subjected to MALDI-TOF for molar mass determination. The resulting peak maximum molar masses were combined with the elution volumes of the fractions from SEC to give a PDMS calibration curve log M vs V_e. The analysis of random copolyesters has been described recently by Montaudo et al. [46, 91].

A summary of applications where MALDI-TOF mass spectrometry has been used as an off-line detector for liquid chromatography is given in Table 7.7.

Table 7.7. Applications of MALDI-TOF MS as an off-line detector in liquid chromatography

Polymer	LC separation technique	Remarks	Reference
Poly(butylene adipate) and copolyesters	SEC	Calibration of SEC, structural characterization	[31, 51, 62]
Poly(hexene adipate)	LC-CC	Functionality type separation, oligomer distributions	[47]
Aliphatic polyesters	SEC	Molar mass determination	[57]
Polybutyl acrylate	SEC	Calibration of SEC	[61]
Polymethyl methacrylate	SEC	Calibration of SEC	[62]
Polycarbonate	SEC	Calibration of SEC	[61]
Polyethylene oxide	LC-CC	Functionality type separation, oligomer distributions	[43]
Polylactide	LC-CC LC-CC	Functionality type separation, oligomer distributions	[43, 55]
Polymethyl methacrylate	SEC	Functionality type separation, oligomer distributions	43]
Cyclic polyimides	SEC	Identification of SEC fractions, molar mass determination	[48]
Poly(tetrahydropyrene)	Isocratic HPLC, gradient HPLC	Molar mass determination	[49]

Table 7.7. (continued)

Polymer	LC separation technique	Remarks	Reference
Polystyrene macromonomers	Isocratic HPLC, gradient HPLC	Chemical composition of coupling products, molar mass determination	[50]
Carboxy-terminated polystyrene	Gradient HPLC	Identification and chemical composition of fractions	[53]
Macrocyclic polystyrenes	LC-CC	Identification and chemical composition of fractions	[54]
Poly(dimethyl siloxane)	SEC	Calibration of SEC	[52,58,62]
Poly(dimethyl siloxane)	HPLC	Calibration of HPLC, identification	[60]
Poly(dimethyl siloxane)	SEC	Determination of different oligomer distributions	[92]
Epoxy copolymers	SEC	Molar mass determination, chemical composition analysis	[56]
Oligolactones	SFC	Identification, chemical composition of oligomers	[59]
Poly(alkyl thiophenes)	SEC	Molar masses, endgroup structures, modification of endgroups	[96]
PMMA-PBA copolymers	SEC	Analysis of bivariate distributions by SEC/NMR and SEC/MALDI	[97]
Styrene-maleic anhydride copolymers	SEC	Analysis of bivariate distributions by SEC/NMR and SEC/MALDI	[98]
Polyamide-6	LC-CC	Analysis of linear and cyclic oligomers	[99]

7.4 Interfaces for Coupling LC and MALDI-TOF MS

As was already pointed out in Sect. 7.2, the problems of coupling LC and MALDI-TOF MS are related to the fact that MALDI-TOF is based on the pulsed desorption of molecules from a solid surface layer and, therefore, a priori not compatible with liquid chromatography. An alternative is off-line LC separations and subsequent MALDI-TOF analysis of the resulting fractions. Although this is laborious, it has the advantage that virtually any type of chromatographic separation can be combined with MALDI-TOF.

While direct coupling is a problem, it has been shown in recent years that at least the LC sample collection and subsequent preparation of the samples for MALDI-TOF analysis can be automated. Different interfaces have been developed that can be coupled directly to the liquid chromatograph. In these interfaces, the eluate stream is deposited on the MALDI target via a spray or a drip process. The matrix required for the MALDI process is either co-added to the eluate stream or matrix-precoated MALDI targets are used.

7.4.1 LC Transform Interface

About ten years ago Lab Connections Inc., U.S.A., introduced a device for interfacing liquid chromatography and FTIR spectroscopy. This "LC-Transform" interface, based on the invention of Gagel and Biemann, is composed of two independent modules, the sample collection module and the optics module [63–68]. The effluent of the liquid chromatography column is split with a fraction (frequently 10% of the total effluent) going into the heated nebulizer nozzle located above a rotating sample collection disc. The nozzle rapidly evaporates the mobile phase while depositing a tightly focused track of the solute. When a chromatogram has been collected on the sample collector disc, the disc is transferred to the optics module in the FTIR for analysis of the deposited sample track. A control module defines the sample collection disc position and rotation rate in order to be compatible with the run time and peak resolution of the chromatographic separation.

One of the first applications of an LC-Transform interface for collecting SEC fractions has been described by Kassis et al. [93]. The sample collector disc has been precoated with an appropriate matrix. Molar mass information has been obtained for fractions of PMMA. The FTIR interface has been modified such that instead of a rotating germanium disc a MALDI target, mounted on a programmable X-Y table, is coated [94]. The principal construction of the spray nozzle is schematically presented in Fig. 7.26A; a view on the LC-Transform Model 600 is given in Fig. 7.26B. Similar to the FTIR interface, the eluate from the LC separation enters the heated nebulizer capillary where it is mixed with the sheath gas. An aerosol is formed which is sprayed on the moving sample target. During the spraying process the mobile phase is evaporated and a solid layer of material is formed on the MALDI target. The matrix required for the MALDI process can be introduced in two different ways. The simplest way is to precoat the sample target with the appropriate matrix. Lab Connections Inc. offers one-time

7.4 Interfaces for Coupling LC and MALDI-TOF MS

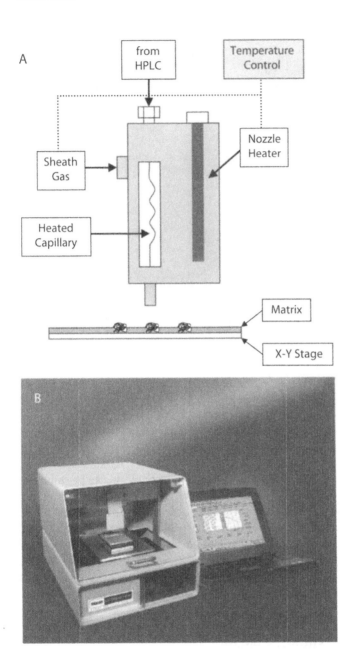

Fig. 7.26. A Schematic representation of the spray nozzle; B a view of the Model 600 LC-Transform (courtesy of Lab Connections Inc.)

foils that are precoated with a variety of matrices and can be fixed to the MALDI target via a conductive adhesive backing that sticks to the target; see Fig. 7.27. Another option is to add the matrix solution to the LC eluate prior to entering the nebulizer nozzle.

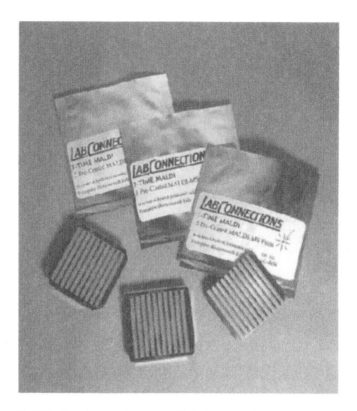

Fig. 7.27. One-time matrix-precoated foils (courtesy of Lab Connections Inc.)

Using this experimental setup SEC separation can be combined with MALDI-TOF analysis. In the following application of Lab Connections Inc. two polyethylene glycols having molar masses of 4600 g/mol and 1500 g/mol, respectively, are separated by SEC into four fractions A–D and analyzed by MALDI-TOF; see Fig. 7.28. For all fractions well-resolved spectra are obtained that can be used for further analysis of the chemical composition and oligomer distribution of the fractions.

7.4.2 PROBOT Interface

With the growing interest in microcolumn liquid chromatography, such as micro-, capillary-, and nanoscale HPLC, as well as capillary electrophoresis, more and more dedicated systems became commercially available. Today there are no problems in running microcolumn HPLC separations with sufficient precision, reproducibility, and sensitivity. The low flow rates typically in the

7.4 Interfaces for Coupling LC and MALDI-TOF MS

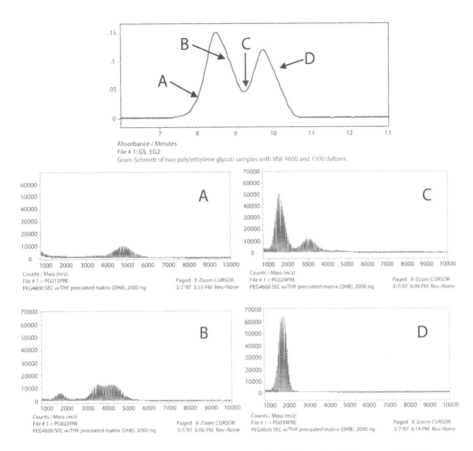

Fig. 7.28. SEC-MALDI-TOF analysis of a mixture of two polyethylene glycols (courtesy of Lab Connections Inc.)

magnitude of a few µl/min allow for microfractionation directly onto targets for MALDI-TOF MS. These low flow rates assure precise deposition of the microcolumn eluate with minimal dispersion and without any risk of flooding the targets. By using a robotic system with precise X-Y-Z movements with a high resolution (±2.5 µm) fully automated microfraction collection becomes possible.

For interfacing microcolumn HPLC or capillary electrophoresis with MALDI-TOF MS, Bioanalytical Instruments (BAI), Germany, developed the robotic interface PROBOT [69,70], see schematic representation in Fig. 7.29. The outlet of the HPLC/CE system is connected to a double-wall stainless steel needle which is positioned above the MALDI target. Via a T-piece matrix solution is coaxially added to the HPLC effluent. The effluent can be placed

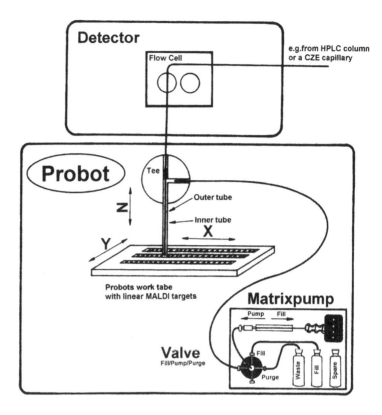

Fig. 7.29. View and schematic representation of the PROBOT interface (courtesy of BAI, Germany)

dropwise or as a continuous track on the MALDI target by moving the X-Y table. After evaporating the solvent, discrete or continuous spots of sample fractions and matrix are formed which can be analyzed directly by MALDI-TOF [71].

7.5 Measurement of LC Fractions Using a MALDI-TOF Interface

Similar to the off-line measurements decribed in Sect. 7.3, the "on-line" setup via an interface can be used to analyze fractions from a variety of different chromatographic separations, including SEC, HPLC, and LC-CC. The following applications present typical examples using either the LC-Transform or the PROBOT interface.

7.5.1 SEC-MALDI-TOF Analysis of Poly(n-butyl methacrylate-*block*-methyl methacrylate) [72]

Aim

Block copolymers are frequently prepared by sequential polymerization using different techniques including anionic, group transfer (GTP), or atom transfer radical polymerization (ATRP). The polymerization is started by homopolymerizing the first monomer to form the first block of the block copolymer. When the first monomer is consumed, the second monomer is added to the reaction mixture to form the second block. Due to impurities in the reaction mixture, very frequently after the first polymerization step a number of chain ends are terminated and a homopolymer fraction of the first monomer is formed.

As has been discussed in Sect. 7.3.1, accurate molar mass determination of copolymers is a problem due to the fact that the SEC system cannot be calibrated by proper calibration standards. Therefore, a calibration shall be conducted through SEC-MALDI-TOF. In addition, information shall be obtained on the presence of homopolymers and the chemical composition of the copolymer fractions.

Materials

Diblock copolymer of n-butyl methacrylate and methyl methacrylate. The sample under investigation has been prepared by group transfer polymerization starting with n-butyl methacrylate. The average molar mass of the sample is about 4000 g/mol.

Equipment

Chromatographic System — modular HPLC system comprising a Waters model 510 pump, a Rheodyne six-port injection valve and a Waters column oven. For automatic fraction collection and deposition on the MALDI targets the LC-Transform Model 500 (Lab Connections, Marlborough, MA, U.S.A.) is used.

Columns — set of four PLgel columns 10^5 Å + Mixed-D + Mixed-E + 50 Å average pore diameter. Column size was 300×7.5 mm I.D.

Mobile Phase — THF of HPLC grade.

Detector — Waters differential refractometer R 410.

Sample Amount — 100 µl of a 0.2 mg/ml solution in the mobile phase.

MALDI-TOF system — Kratos Kompact MALDI 4 with linear and reflectron detectors, acceleration voltage of positive 20 kV.

MALDI-TOF sample preparation — 10% of the SEC effluent were introduced into the interface via a capillary. The effluent was sprayed via a heated capillary nozzle continuously on a slowly moving Kratos MALDI target precoated with the matrix 1,8,9-trihydroxy anthracene (dithranol). The matrix was manually deposited on the MALDI target from a THF solution. For the enhancement of ion production, sodium trifluoroacetate was added to the matrix.

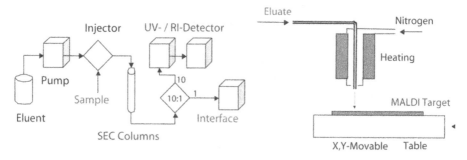

Fig. 7.30. **A** Schematic representation of the experimental setup for the SEC to MALDI-TOF coupling. **B** Deposition of the fractions onto the MALDI target

Preparatory investigations

The experimental setup for the SEC-MALDI-TOF analysis is schematically presented in Fig. 7.30. As was already explained, the SEC separation is carried out in the usual way with typical SEC flow rates and concentration. Then 10% of the effluent is split off, directed into the interface, and deposited on the MALDI targets. The SEC chromatogram of the sample recorded with the refractive index detector does not indicate any peculiarities or by-products; see Fig. 7.31.

MALDI-TOF analysis

After the SEC separation the fractions are automatically deposited on the MALDI-TOF sample target. Prior to fraction deposition the target was precoated with the matrix dithranol and a small amount of sodium trifluoro acetate (NaTFA) to enhance the formation of $[M+Na]^+$ molecular ions. Since the fraction deposition is carried out through a heated capillary nozzle, a solid fraction/matrix film is obtained on the MALDI-TOF target. The spray-deposition procedure must be optimized very carefully in order to assure that a uniform sample/matrix track is formed. If the nozzle temperature is too low, the aerosol stream is too wet, resulting in partial desolving and blowing away of the matrix. If the nozzle temperature is too high, the aerosol stream is too dry, resulting in a bilayer film where the matrix and the sample molecules are not mixed at all. In the present experiments where solely THF was used as the eluent, a spray nozzle temperature of 70°C has been found to be the optimum. In the same manner, the distance between the spray nozzle and the MALDI-TOF target has to be optimized.

The MALDI-TOF target (Kratos) has a length of 70 mm and is scanned continuously with 3500 laser pulses. Every 50 pulses are summarized to give a complete MALDI-TOF spectrum. With SEC as the preseparation technique, low positions on the target correspond to high molar masses, while high positions are equivalent to low molar masses. Selected spectra from different positions of

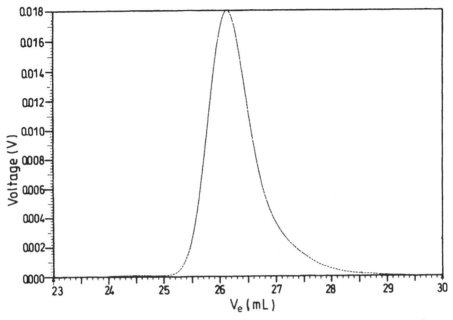

Fig. 7.31. SEC chromatogram of the PnBMA-PMMA block copolymer; stationary phase: PLgel 10^5 Å + Mixed-D + Mixed-E + 50 Å, mobile phase: THF, detector: RI

the polymer/matrix track of the sample are given in Fig. 7.32. The higher molar mass fractions in (A)–(C) are characteristic of copolymer structures exhibiting typical mass increments of 100 Da for the MMA repeat unit and 142 Da for the nBMA repeat unit. Even these narrow disperse fractions exhibit a multitude of different mass peaks (usually more than 100) indicating the high complexity of the fractions. Different from A–C, the lower molar mass fraction in (D) is very uniform with respect to composition. For this fraction, only peak-to-peak mass increments of 142 Da are observed which are typical for PnBMA. Accordingly, this fraction can be assigned to an unwanted by-product, namely butyl methacrylate homopolymer.

The total spectrum of the sample is obtained when the individual spectra of all pulses are summed; see Fig. 7.33A. This spectrum clearly indicates the presence of PnBMA having an average molar mass which is about 50% of the total molar mass of 4200 g/mol. This result is in very good agreement with the expectations. The monomer ratio of the sample under investigation is about 1:1 resulting in average block lengths of 2100 g/mol for the PnBMA and the PMMA blocks. Since the polymerization was started with n-butyl methacrylate, the formation of a small amount of PnBMA has to be expected. The molar mass of this homopolymer must be

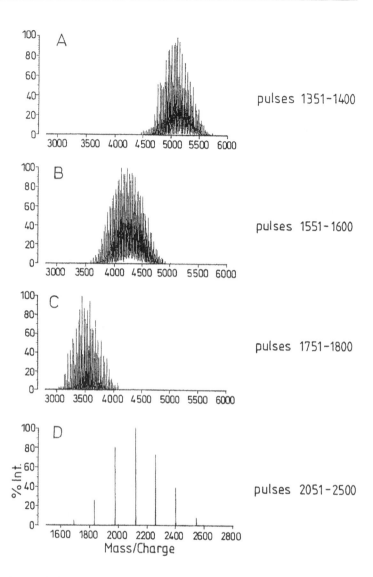

Fig. 7.32A–D. MALDI-TOF spectra of fractions from SEC separation of PnBMA-block-PMMA, on-line SEC-MALDI-TOF analysis, matrix: dithranol, NaTFA. (Reprinted from [72] with permission of Elsevier Science Ltd., UK)

of the same magnitude as the PnBMA block in the copolymer. This is indeed the case as is shown by the MALDI-TOF measurement.

The chemical composition of the block copolymer can be studied in detail by analyzing the different mass peaks, see zoomed part

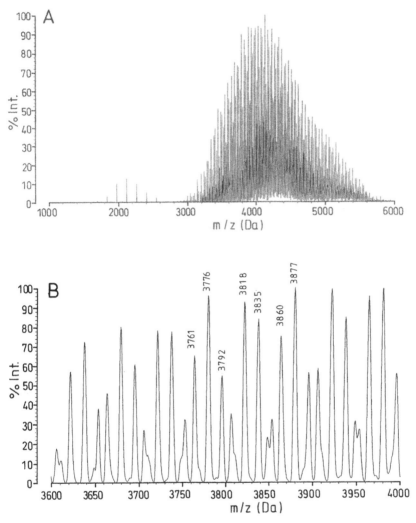

Fig. 7.33. A Calculated MALDI-TOF spectrum from on-line SEC-MALDI-TOF analysis of PnBMA-block-PMMA. **B** Zoomed part of the spectrum. (Reprinted from [72] with permission of Elsevier Science Ltd., UK)

of the spectrum in Fig. 7.33B. Each peak in the spectrum can be assigned to one individual oligomer composition $(nBMA)_X(MMA)_Y$. For example, the mass peak at 3761 Da corresponds to an oligomer with 15 nBMA and 16 MMA units. Its gross structure is H-$(nBMA)_{15}$-$(MMA)_{16}$-H. The mass peak at 3835 Da is due to the oligomer H-$(nBMA)_{12}$-$(MMA)_{21}$-H. The calculated and observed molar masses for selected oligomers are summarized in Table 7.8.

Table 7.8. Calculated and observed molar masses for individual oligomers of sample PnBMA-*b*-PMMA

$$H\text{---}\left[\text{---}CH_2\text{---}\underset{\underset{COOC_4H_9}{|}}{\overset{\overset{CH_3}{|}}{C}}\text{---}\right]_X\left[\text{---}CH_2\text{---}\underset{\underset{COOCH_3}{|}}{\overset{\overset{CH_3}{|}}{C}}\text{---}\right]_Y\text{---}H$$

X	Y	[M+Na]$^+$ (calculated)	[M+Na]$^+$ (observed)
15	16	3760	3761
13	19	3776	3776
11	22	3792	3792
16	15	3802	3803
14	18	3818	3818
12	21	3834	3835
10	24	3850	3850
15	17	3860	3860
13	20	3876	3877

7.5.2 Analysis of Calixarenes by LC-CC-MALDI-TOF [73]

Aim Calixarenes are cyclic oligomeric compounds which are formed in the reaction of phenols and formaldehyde. Calixarenes were discovered for the first time as by-products in the production of phenolic novolacs. They are excellent host molecules for anions, cations, and neutral molecules and are used in medical diagnostics, sensor technology, and waste water treatment.

Very frequently it is of interest to determine calixarenes in linear novolacs or to separate them according to their substituents irrespective of their ring size. The present application describes a separation procedure and the on-line analysis of the chromatographic fractions by MALDI-TOF.

Materials Calixarenes of different ring size and different substituents at the phenolic nuclei are commercially available [74,75].

Equipment Chromatographic system — Hewlett Packard HP 1090 HPLC system. For automatic fraction collection and deposition on the MALDI targets the LC-Transform Model 500 (Lab Connections, Marlborough, MA, U.S.A.) is used.

Columns — YMC ODS, 120 Å average pore diameter. Column size was 250×4.6 mm I.D.

Mobile phase — mixtures of THF and water (+0.1% trifluoro acetic acid), all of HPLC grade.

Detector — evaporative light scattering detector ERC Sedex 45.

Sample amount — 10 µl of a 15 mg/ml solution in the mobile phase.

7.5 Measurement of LC Fractions Using a MALDI-TOF Interface

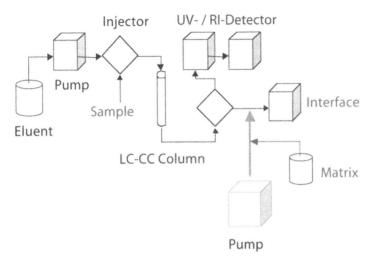

Fig. 7.34. Schematic representation of the experimental setup for the LC-CC to MALDI-TOF coupling using coaddition of the matrix

MALDI-TOF system — Kratos Kompact MALDI 3 with linear and reflectron detectors, acceleration voltage of positive 20 kV.

MALDI-TOF sample preparation — 10% of the SEC effluent were introduced into the interface via a capillary. The effluent was sprayed via a heated capillary nozzle continuously on a slowly moving Kratos MALDI target. Before entering the spray nozzle, a matrix solution of 2,4,6-trihydroxy acetophenone in THF was continuously coadded to the eluate stream. The matrix flow was 0.2 ml/min while the eluate flow was 0.5 ml/min.

Preparatory Investigations

The experimental setup for the LC-CC-MALDI-TOF analysis is schematically presented in Fig. 7.34. Different from the previous application, the MALDI targets are not precoated. Instead, the matrix solution is coadded to the eluate before entering the LC transform interface. This premixing of the sample and the matrix produces an ideal coverage of the sample molecules by matrix molecules. By optimizing the gas flow and the spray nozzle temperature conditions must be found that prevent the matrix from crystallizing at the tip of the spray nozzle.

For separating the calixarenes with respect to chemical composition and regardless of ring size, the critical eluent composition must be determined. This is done by measuring calixarenes of different sizes at eluents of different composition, see also sections 7.3.3 and 7.3.5. For p-octyl calixarenes the critical point of adsorption corresponds to a mobile phase of THF/water (+0.1% TFA) 91.5:8.5% by weight. At this eluent composition, p-octyl calixarenes of different ring size elute at the same retention time; see

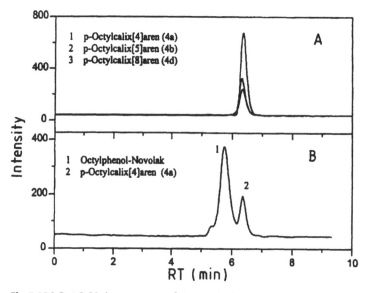

Fig. 7.35 A,B. LC-CC chromatogram of: **A** p-octyl calixarenes; **B** their mixture with a linear p-octylphenol novolac; stationary phase: YMC ODS, mobile phase: THF-water (+0.1% TFA) 91.5:8.5, detector: ELSD. (Reprinted from [73] with permission of GIT Verlag, Germany)

MALDI-TOF analysis

Fig. 7.35A. Under the same conditions, linear p-octylphenol novolacs and cyclic oligomers can be separated; see Fig. 7.35B.

The MALDI-TOF analysis of a mixture of the linear and the cyclic fraction of an octylphenol novolac separated by LC-CC is presented in Fig. 7.36. The most intense elution peak can be identified as the linear novolac structure; see Pos. 5 and 6, with the following chemical structure:

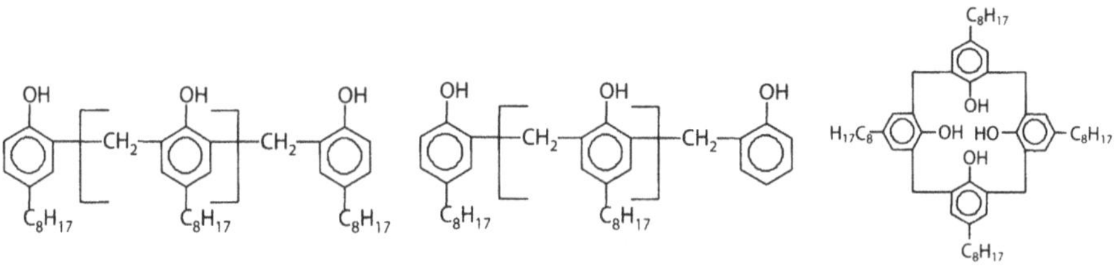

The small peak eluting at lower retention times, see Pos. 1, can be assigned to a linear novolac structure **2** where one p-octylphenol is replaced by a phenol. In Pos. 3 an overlapping of both distributions is obtained. The cyclic oligomer elutes after the linear oligomers, as can be identified by the MALDI spectrum in Pos. 9. The mass peak at 895 Da corresponds to the $[M+Na]^+$ molecular ion of the calixarene **3**.

7.5 Measurement of LC Fractions Using a MALDI-TOF Interface 265

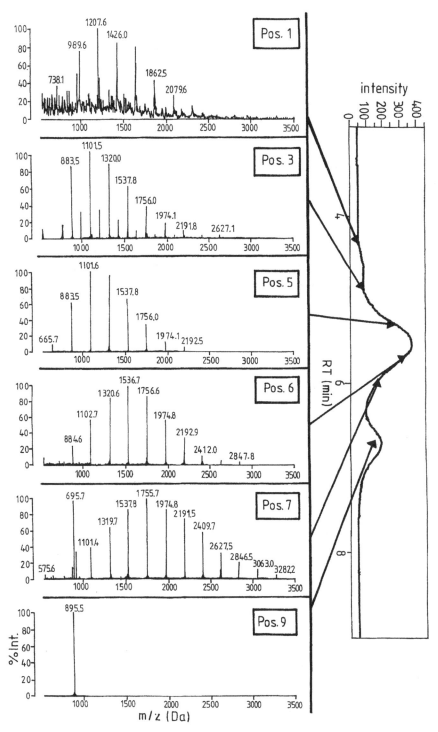

Fig. 7.36. Selected MALDI-TOF spectra of different fractions obtained by the LC-CC separation presented in Fig. 7.35 B. (Reprinted from [73] with permission of GIT Verlag, Germany)

7.5.3 Analysis of Polycarbonate by µSEC-MALDI-TOF [76, 77]

Aim Apart from the spray deposition interface described in Sect. 7.5.1 and 7.5.2, liquid chromatographic fractions can be deposited on a MALDI-TOF target dropwise. This is particularly useful when very low flow rates are used in the LC separation. The following application uses µSEC with microcolumns and a flow rate of 5 µl/min to separate polycarbonate. The resulting fractions are used to calibrate the SEC system.

Materials Commercial bisphenol A polycarbonate with an average molar mass of M_w 28,800 g/mol.

Equipment Chromatographic system — applied Biosystems model 140B HPLC pump equipped with a Valco Cheminert model C4-1004-0.5 injection valve with an internal loop of 500 nl. For automatic fraction collection and deposition on the MALDI targets the Probot interface (BAI, Lautertal, Germany) is used.

Columns — µSEC column packed with PL-gel MiniMixed-D, 30 cm × 800 µm i.d., 5 µm average particle size.

Mobile phase — THF stabilized with 0.025% BHT.

Detector — Bischoff Lambda 1000 UV absorbance detector with capillary flow cell (Prince Technologies, U.S.A.).

Sample amount — 500 nl of a 3 mg/ml solution in THF

MALDI-TOF system — Bruker Biflex with linear and reflectron detectors, delayed extraction Scout ion source and Himass detector in the linear mode.

MALDI-TOF sample preparation — the SEC effluent was directly spotted on the MALDI target via the Probot interface. The matrix solution was coaxially added to the eluate stream via a BAS Baby Bee syringe pump operated at 2.5 µl/min. As the matrix *trans*-indoleacrylic acid was used.

Preparatory investigations The experimental setup for the µSEC-MALDI-TOF analysis is schematically presented in Fig. 7.37.

The polycarbonate sample is dissolved in THF and injected into the µSEC system. The chromatographic polymer distribution thus obtained is automatically fractionated as follows: the fused silica exit capillary is inserted through the T-piece of the robotic interface to the tip of the deposition needle; see Fig. 7.37. At the needle tip, the µSEC effluent is coaxially mixed with a fixed concentration and flow rate of the matrix solution and spotted onto the MALDI targets. Thus, 32 spots are obtained for MALDI analysis, each corresponding with a 10 s elution window and containing a narrow polymer distribution. Typically, seven or eight spots are selected, equally spread over the chromatographic elution range, for absolute SEC calibration through m/z determination via

7.5 Measurement of LC Fractions Using a MALDI-TOF Interface

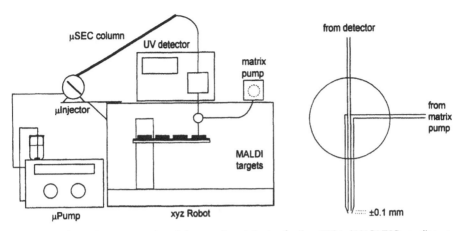

Fig. 7.37. Schematic representation of the experimental setup for the μSEC to MALDI-TOF coupling using the Probot interface. (Reprinted from [76] with permission of the American Chemical Society, U.S.A.)

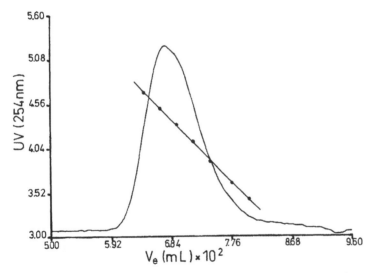

Fig. 7.38. μSEC chromatogram of the polycarbonate; stationary phase: PL-gel MiniMixed-D, mobile phase: THF, detector: UV. (Reprinted from [76] with permission of the American Chemical Society, U.S.A.)

MALDI-TOF. The μSEC chromatogram of the polycarbonate and the MALDI-TOF based calibration curve are presented in Fig. 7.38.

The MALDI-TOF spectra taken from different spots (fractions 4, 10, 16, 19) are presented in Fig. 7.39. They are acquired in the continuous extraction positive ion linear mode and are the sum of 100 laser shots. The summed mass spectra (ion intensity vs m/z) are converted into [(intensity × m/z) vs log m/z]. The top of the

MALDI-TOF analysis

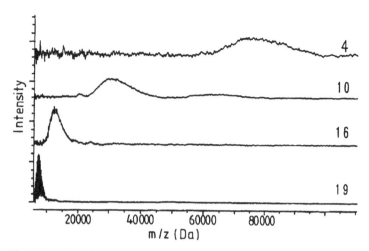

Fig. 7.39. Selected MALDI-TOF spectra of different fractions obtained by the µSEC separation presented in Fig. 7.38. (Reprinted from [76] with permission of the American Chemical Society, U.S.A.)

Table 7.9. Molar mass data of the polycarbonate determined by µSEC-MALDI-TOF and other methods (in g/mol)

Method	M_w	M_n	M_w/M_n
Manufacturer data [76]	28,800	17,300	1.66
µSEC-MALDI-TOF	26,400	15,800	1.67
MALDI-TOF MS	11,500	9,600	1.20
SEC with PS calibration	53,100	28,500	1.86

weight fraction distribution, M_p, is used as the absolute mass value in the calibration of the µSEC system.

The absolute molar mass data are summarized in Table 7.9 and compared with polystyrene-based data and data obtained by MALDI-TOF without prefractionation. From these data it can be concluded that both direct MALDI-TOF MS and SEC alone (with PS calibration) yield incorrect results, while the µSEC-MALDI-TOF results are in fairly good agreement with the manufacturer's data.

In addition to SEC calibration, the MALDI-TOF spectra yield structural information. In particular, isotopically resolved oligomer peaks obtained in the delayed extraction reflectron mode can be used to identify different endgroups. A detail of the mass spectrum of spot 22 (corresponding to an elution volume of 77 µl) is shown in Fig. 7.40. The $[M+Na]^+$ and $[M+K]^+$ molecular ions of bisphenyl-terminated oligomers can be identified at 2271 and

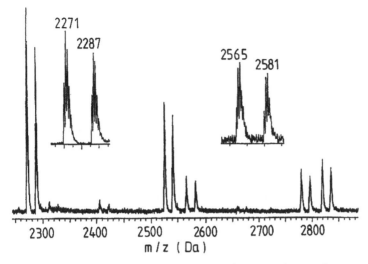

Fig. 7.40. Detail of the mass spectrum of spot 22 of the polycarbonate after µSEC-MALDI-TOF MS. (Reprinted from [76] with permission of the American Chemical Society, U.S.A.)

2287 Da. Other oligomer series correspond to monophenyl-terminated and cyclic oligomers [78].

7.5.4 SEC-MALDI-TOF and LC-CC-MALDI-TOF of Polypropylene Oxides [79]

Polymers and copolymers of propylene oxide (PPO) are in widespread use in technical applications. The copolymerization of ethylene oxide and propylene oxide yields segmented polymers with amphiphilic properties which are used as surfactants. Glycerol-started PPO is frequently used as a cross-linking polyol for the production of polyurethanes.

In addition to the molar mass distribution (MMD) PPOs are distributed with respect to functional endgroups. This functionality type distribution (FTD) results from the presence of water or other impurities in the reaction mixture. For a proper evaluation of the product properties of PPOs it is, therefore, important to determine FTD as a function of molar mass.

In the following application, the analysis of functionally heterogeneous PPOs by LC-CC shall be discussed. As the liquid chromatographic separation techniques SEC and LC-CC shall be used to demonstrate the differences in the information obtainable from both techniques.

Aim

Materials

The PPO samples are a commercial products of Arcol Co., U.S.A. Samples 1 and 2 have average molar masses of 1500 g/mol and 8500 g/mol, respectively.

Equipment

Chromatographic system — Waters Model 510 HPLC pump with a Rheodyne six-port injection valve. For automatic fraction collection and deposition on the MALDI targets the Probot interface (BAI, Lautertal, Germany) is used.

Columns — SEC: PL-gel Mixed-E, 300×7.8 mm i.d. LC-CC: Macherey-Nagel Nucleosil 5 NH_2, 200×4 mm i.d.

Mobile phase — SEC: THF, LC-CC: hexane-isopropanol 92:8 by volume.

Detector — Waters Model 410 refractive index detector or Altech Model 500 ELSD detector.

Sample Amount — 100 µl of a 1 mg/ml (SEC) or 20 µl of a 10 mg/mL (LC-CC) solution in the mobile phase

MALDI-TOF system — Kratos Kompact MALDI 4 with linear and reflectron detectors, acceleration voltage of positive 20 kV.

MALDI-TOF sample preparation — the SEC effluent was directly spotted on the MALDI target via the Probot interface. The matrix solution was coaxially added to the eluate stream via a BAS Baby Bee syringe pump. As the matrix dithranol+LiTFA was used.

Preparatory Investigations

The chromatographic separations are carried out using standard equipment, a refractive index detector or an ELSD is applied for monitoring the separation. The flow rate is 0.5 ml/min. After the chromatography the eluate stream is split with 40 parts of the eluate going to the detector and 1 part going to the Probot interface. With an eluate flow of 12.5 µl/min reaching the interface, the dual collection mode is used to deposit 2 µl/min of the eluate on the MALDI-TOF target. The matrix is coaxially added to the eluate stream before deposition. After the evaporation of the solvent, the MALDI-TOF target is inserted into the spectrometer and spectra are taken from all positions using an automatic scan mode.

In the following experiments, glycerol-started polyproylene oxides are analyzed by liquid chromatography and MALDI-TOF mass spectrometry. The idealized chemical structure of glycerol-started PPO is the following:

$$\begin{array}{l} CH_2-O-[CH_2CH(CH_3)O]_n-H \\ | \\ CH-O-[CH_2CH(CH_3)O]_n-H \\ | \\ CH_2-O-[CH_2CH(CH_3)O]_n-H \end{array}$$

It is assumed that all three hydroxy groups of the glycerol are roughly equally reactive and, therefore, the length distribution of the PPO polymer chains bound to the primary and secondary hy-

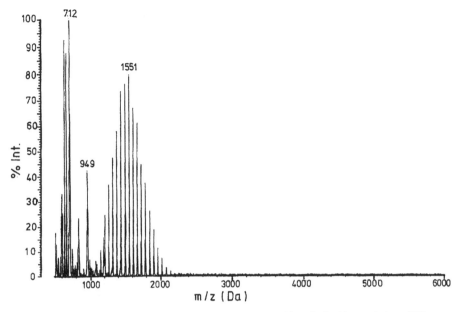

Fig. 7.41. Off-line MALDI-TOF spectrum of sample 1. (Reprinted from [79] with permission of Wiley-VCH, Germany)

droxy groups are similar. Due to the presence of small amounts of water in the reaction mixture, very frequently small amounts of polypropylene glycol are formed in addition to the PPO-triol:

$$H_2O + CH_2\underset{O}{-}CH-CH_3 \longrightarrow HO-[CH_2CH(CH_3)O]_n-H$$

The MALDI-TOF spectrum of the PPO-triol 1 is presented in Fig. 7.41.

The average molar mass of the sample is 1500 g/mol which is in very good agreement with the most intense mass peak at 1551 Da, found in the spectrum. The spectrum exhibits one oligomer series which can be assigned to the PPO-triol structure. Since LiTFA has been added to the matrix, cationization takes place via the attachment of Li^+ to form $[M+Li]^+$ molecular ions:

$[M+Li]^+$ = 6.94 (Li) + 92.10 (Glycerol) + 58.08 (PO) × n

n	$[M+Li]^+$ (calc.)	$[M+Li]^+$ (exp.)
23	1434.88	1435
24	1492.96	1493
25	1551.04	1551

MALDI-TOF analysis

The present spectrum does not show any other oligomer series and, therefore, does not indicate other functionality fractions.

In a second step, the sample is separated with respect to molar mass by SEC and different fractions are subjected to MALDI-TOF mass spectrometry via automated fraction collection and deposition on the MALDI-TOF target. The SEC chromatogram and three selected mass spectra are presented in Fig. 7.42.

Fraction A contains a high molar mass portion of PPO-triol which has not been detected by the off-line MALDI-TOF experiment. Fraction B is also composed of PPO-triol and exhibits three molar mass fractions centered at 4500, 2800, and 1800 Da. The MALDI-TOF spectrum of the fraction C (not shown here) is very similar to the off-line spectrum in Fig. 7.41 and contains the major amount of the PPO-triols. Finally, the lowest molar mass fraction D shows, in addition to the oligomer series of the PPO-triols, a second oligomer series centered at 950 Da (indicated by arrow). This oligomer series can be assigned to a small portion of polypropylene glycol (PPO-diol) which has not been detected in the off-line MALDI experiment. Due to the overlapping with the PPO-triol oligomer series, the PPG peaks can barely be recognized. The peak masses of this series correspond to $[M+Li]^+ = 25+58n$, where n is the degree of polymerization.

Much more selectively than SEC, LC-CC can be used to obtain information on the structural peculiarities of the sample. LC-CC separates functionally heterogeneous samples with regard to their functional groups irrespective of the molar mass. For this type of experiments, in a first step the critical conditions for the respective homopolymer, i.e., PPO, must be determined. This can be done by running PPGs with different molar masses in binary eluents of varying composition. In the present case, the critical point of adsorption for PPG is obtained at an eluent composition of *n*-hexane/iso-propanol 92:8 by volume. The functionality type separation of sample 1 under these conditions is shown in Fig. 7.43. The LC-CC chromatogram exhibits three peaks at elution volumes of 3.3 ml, 5.1 ml, and 8.2 ml (the peak at 2.4 ml is the injection peak).

Similar to the SEC experiment, the LC-CC elution peaks are analyzed by direct coupling to MALDI-TOF MS. While for the first fraction a spectrum cannot be obtained due to very low concentration, well resolved spectra for peaks B and C are shown in Fig. 7.43. Different from SEC, the LC-CC peaks are uniform with respect to chemical composition (functionality). Peak B can be identified as the diol (PPG) fraction, while peak C is due to the PPO-triol oligomers. In agreement with the SEC/MALDI-TOF results the PPG fraction is centered at 950 Da; see spectrum B. The

7.5 Measurement of LC Fractions Using a MALDI-TOF Interface 273

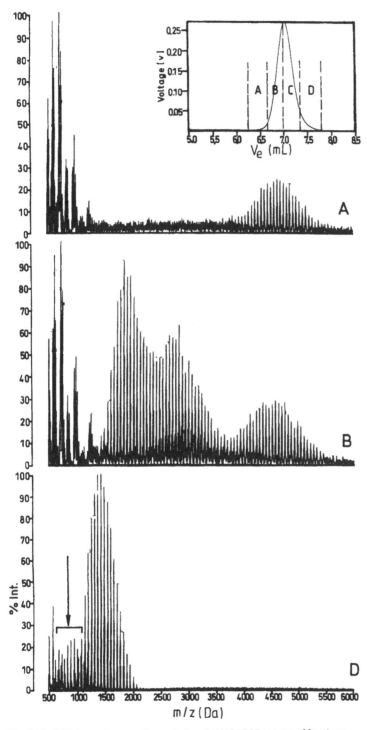

Fig. 7.42. SEC chromatogram of sample 1 and MALDI-TOF spectra of fractions. (Reprinted from [79] with permission of Wiley-VCH, Germany)

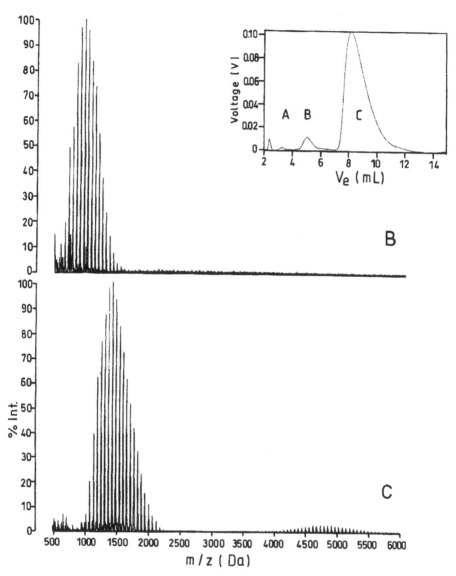

Fig. 7.43. LC-CC chromatogram of sample 1 and MALDI-TOF spectra of fractions. (Reprinted from [79] with permission of Wiley-VCH, Germany)

PPO-triol fraction exhibits a bimodal oligomer distribution, the major distribution being centered at 1500 Da. The second distribution is centered at 4500 Da and is very low in concentration. To summarize, using the combination of LC-CC separation and MALDI-TOF detection it is possible to identify PPG as a by-product in the PPO-triol and to obtain information on the oligomer distribution of both functionality fractions. While the PPG exhib-

7.5 Measurement of LC Fractions Using a MALDI-TOF Interface

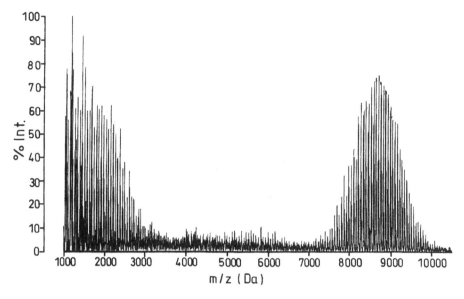

Fig. 7.44. Off-line MALDI-TOF spectrum of sample 2. (Reprinted from [79] with permission of Wiley-VCH, Germany)

its a monomodal oligomer distribution, the PPO-triol is composed of two fractions with different molar masses.

The off-line MALDI-TOF spectrum of a second technical PPO-triol (sample 2) with an average molar mass of 8500 g/mol is presented in Fig. 7.44. Three different molar mass fractions centered around 9000, 6000 and 2000 Da are obtained, the highest molar mass fraction being the PPO-triol.

For a more detailed insight, the sample is fractionated by SEC and fractions A–D analyzed by MALDI-TOF mass spectrometry. Fraction A, C, and D are presented in Fig. 7.45. As is shown in Fig. 7.45A, fraction A exhibits two oligomer distributions. The higher molar mass peak series can be identified as PPO-triol, while the peak series centered around 5500 Da belongs to PPG. In fraction C, a low molar mass fraction of PPO-triol is observed, and in addition an oligomer series appears (indicated by arrow) which cannot be attributed to PPO-triol or PPG. The observed peak series ...2214-2272-2330... can be assigned to PPOs with allylic endgroups or cyclic PPO oligomers:

$$CH_2=CHCH_2O-[CH_2CH(CH_3)O]_n-H$$

Unfortunately, MALDI-TOF mass spectrometry is not able to differentiate the two different structures because their calculated oligomer masses do not deviate from each other.

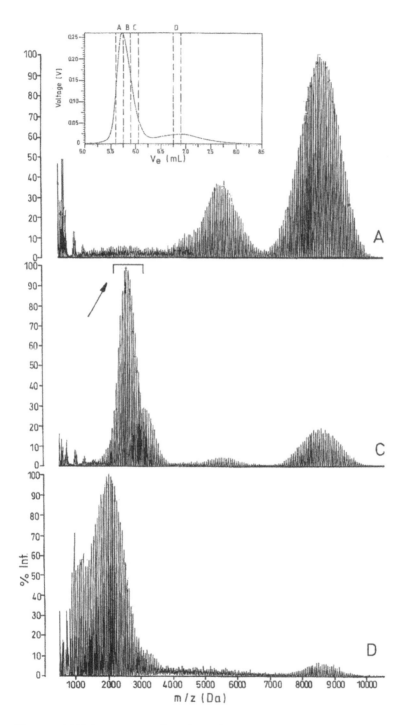

Fig. 7.45. A SEC chromatogram of sample 2 and MALDI-TOF spectra of fractions. (Reprinted from [79] with permission of Wiley-VCH, Germany)

7.5 Measurement of LC Fractions Using a MALDI-TOF Interface   277

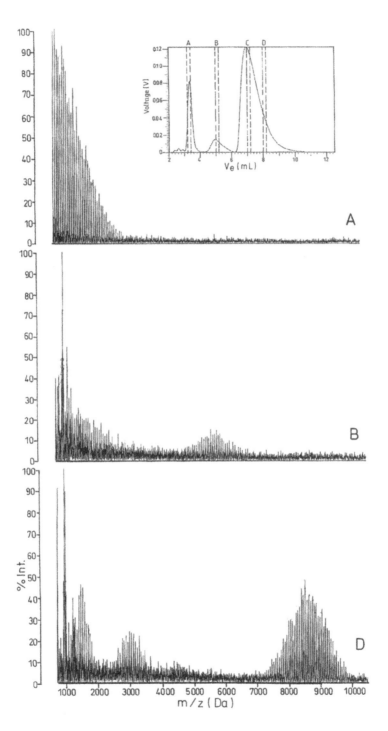

Fig. 7.45. B LC-CC chromatogram of sample 2 and MALDI-TOF spectra of fractions. (Reprinted from [79] with permission of Wiley-VCH, Germany)

	Allylics	Cyclics
	$[M+Li]^+ = 6.94 + 58.08 + 58.08\,n$	$[M+Li]^+ = 6.94 + 58.08\,n$
n	$[M+Li]^+$ (calc.)	$[M+Li]^+$ (calc.)
37	**2213.98**	2155.90
38	2272.06	**2213.98**
39	2330.14	2272.06

Fraction D mainly contains PPOs with allylic endgroups and/or cyclic oligomers.

As has been shown previously, the prefractionation of the sample by LC-CC is the most versatile approach for a complete analysis. The LC-CC chromatogram of sample 2 in Fig. 7.45B indicates that the sample contains three different functionality fractions. In agreement with the results for sample 1, the latest eluting fraction can be assigned to the PPO-triol. The fraction eluting at an elution volume of 5 ml corresponds to PPG. The fraction eluting at the lowest elution volume is significantly less polar than PPO-triol and PPG. This fraction is tentatively assigned to allyl-terminated PPO and cyclic oligomers.

The analysis of the LC-CC fractions by MALDI-TOF reveals the principal complexity of the fractions. The allylic/cyclic fraction A in Fig. 7.45B exhibits one oligomer distribution centered at around 800 Da. The PPG fraction B in Fig. 7.45B is bimodally distributed, one distribution centered at 5600 Da while the other distribution is centered at 1800 Da. Finally, the PPO-triol fraction D in Fig. 7.45B exhibits three oligomer series centered at 8600 Da, 3100 Da, and 1500 Da. The total structural information obtained for sample 2 by coupled LC-CC and MALDI-TOF mass spectrometry is summarized in Table 7.10.

Table 7.10. Analysis of sample 2 by coupled LC-CC and MALDI-TOF MS

LC-CC fraction	A	B	C+D
Structure	$CH_2=CHCH_2O[CH_2CH(CH_3)O]_nH$ $\boxed{\quad -[CH_2CH(CH_3)O]_n - \quad}$	$HO-[CH_2CH(CH_3)O]_n-H$	$CH_2-O-[CH_2CH(CH_3)O]_n-H$ $\mid$ $CH-O-[CH_2CH(CH_3)O]_n-H$ $\mid$ $CH_2-O-[CH_2CH(CH_3)O]_n-H$
	$[M+Li]^+ = 6.94 + 58.08\,n$	$[M+Li]^+ = 24.96 + 58.08\,n$	$[M+Li]^+ = 99.04 + 58.08\,n$
Molar mass oligomer series (M_p)	400–3200 Da (800)	800–3000 Da (1800) 4900–6400 Da (5600)	1300–2000 Da (1500) 2400–3600 Da (3100) 7000–10,000 Da (8600)

7.6 Direct Coupling of Liquid Chromatography and MALDI-TOF MS

Considering the potential of MALDI-TOF in terms of versatility and sensitivity, on-line coupling with liquid chromatography would be a highly attractive possibility. Given the experiences with the direct introduction of small matrix-containing liquid streams into high-vacuum instruments, it took surprisingly long before a device for liquid introduction to MALDI was described.

There are two approaches to direct coupling: continuous flow and aerosol. Continuous-flow (cf) MALDI-TOF uses a flow probe similar to a cf fast atom bombardment probe [80,81]. The maximum cf-MALDI flow rate is less than 5 µl/min; thus only a fraction of the flow from a conventional LC column can be used. There is an additional restriction that the matrix itself be a liquid, such as 3-nitrobenzyl alcohol.

In the aerosol method for MALDI liquid introduction the matrix and analyte are dissolved in a solution that is sprayed directly into the mass spectrometer [82]. In this case, flow rates in the magnitude of 0.5 ml/min can be used. The solvent evaporates when the aerosol passes a heated tube, and ions are formed by pulsed UV laser radiation. Therefore, coupling aerosol MALDI-TOF to liquid chromatography simply involves mixing the column effluent with a matrix prior to nebulization [83].

The experimental setup of an aerosol MALDI-TOF MS in both linear and reflectron configurations has been described by Murray and Fei [84,85]. An aerosol containing both matrix and analyte is produced in vacuum using a pneumatic nebulizer with nitrogen as the carrier gas. The solvent is removed as the particles pass through a heated drying tube, and ions are formed in the gas phase with 10 Hz pulsed 355 nm radiation from a frequency-tripled Nd:YAG laser. Ions are formed in the single-stage acceleration region; the ion energies are approximately 5 kV. The ions are mass separated in a two-stage reflectron TOF instrument and detected by a 40-mm dual-microchannel plate particle multiplier. The coupling of this instrument to SEC is shown in Fig. 7.46.

The SEC eluent is mixed with the matrix and then delivered to the mass spectrometer. In the following application methanol is the mobile phase, the flow rate is 0.2 ml/min. The matrix is ferulic acid acidified with trifluoroacetic acid.

The separation and MALDI-TOF analysis of a polethylene glycol, average molar mass 1000 g/mol, is presented in Fig. 7.47A. A total of 93 spectra are obtained continuously during the SEC separation of PEG 1000. Each spectrum is the average of 100 laser shots. The ionization takes place through protonation and the at-

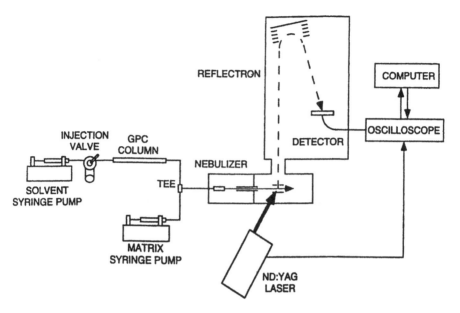

Fig. 7.46. Aerosol MALDI-TOF instrument coupled to SEC. (Reprinted from [86] with permission of American Chemical Society, U.S.A.)

tachment of Na$^+$ cations. Accordingly, two oligomer distributions are obtained in the spectra corresponding to the [M+H]$^+$ and [M+Na]$^+$ molecular ions. Selected ion chromatograms can be obtained by integrating the peak intensity for a particular oligomer in each mass spectrum and plotting the result as a function of retention time; see Fig. 7.47B. The single ion chromatograms can be used to calibrate the SEC system by plotting the molar mass of the respective oligomer vs its retention time.

Although continuous-flow (cf) MALDI-TOF has not been used in polymer analysis yet, its coupling to liquid chromatography shall be reviewed briefly [87–89]. All initial studies on cf-MALDI were done using a linear TOF mass spectrometer where the sample probe was placed in a position orthogonal to the ion flight path. To interface liquid chromatography to cf-MALDI, an on-line post-column matrix addition method has been developed. Figure 7.48 shows the schematic diagram of the LC-MALDI interface (A), including the cf probe (B).

The solvent splitter is used to obtain a flow rate compatible to the micro column LC separation, which is in the range of 1–10 µl/min. For MALDI analysis of the LC fractions, the end of the column is connected to the three-port mixing tee through a short transfer tube. The second port of the tee is connected to a syringe pump that continuously feeds the matrix solution. The resulting

7.6 Direct Coupling of Liquid Chromatography and MALDI-TOF MS

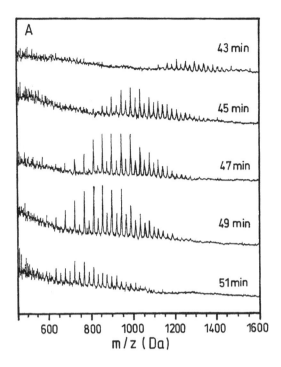

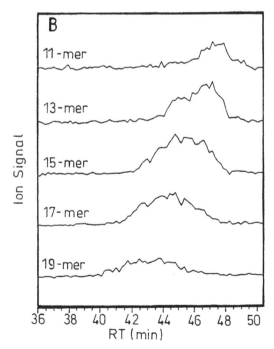

Fig. 7.47. A MALDI-TOF spectra of fractions. **B** Single ion chromatograms obtained from on-line SEC-MALDI-TOF of PEG 1000. (Reprinted from [86] with permission of American Chemical Society, U.S.A.)

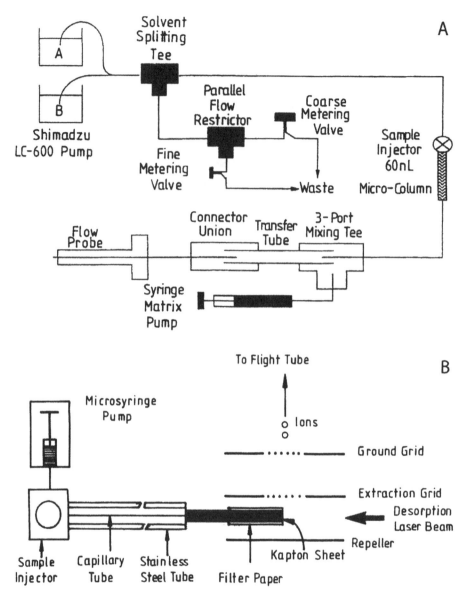

Fig. 7.48. A Schematic construction of the microcolumn LC-MALDI system. **B** Using a cf probe. (Reprinted from [89] with permission of Elsevier Science B.V., The Netherlands)

mixture flows to the flow probe. The flow probe is inserted between the repeller and extraction plates of a linear TOF MS where a 266 nm UV laser beam from a Nd:YAG laser generates the ions.

Using this experimental setup protein and peptide mixtures have been separated by reversed-phase chromatography and the MALDI spectra of the fractions were taken. While the ion chroma-

7.6 Direct Coupling of Liquid Chromatography and MALDI-TOF MS

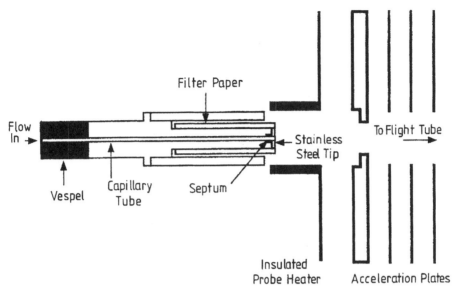

Fig. 7.49. Schematic construction of the cf-MALDI probe used for parallel ion extraction. (Reprinted from [89] with permission of Elsevier Science B.V., The Netherlands)

tograms were very similar to the corresponding UV chromatograms, the mass resolution of the MALDI measurements was very low, about 10–20 fwhm (full width half maximum). The reason for the low resolution was the orthogonal arrangement of the laser beam and the flight tube. Therefore, subsequent research was focused on improving the resolution of the system. This has been achieved by developing a cf-MALDI probe with a parallel ion extraction geometry; see Fig. 7.49.

The cf probe is inserted into the center of the first acceleration plate. With parallel extraction cf-MALDI, stable flow, and reproducible analyte signals can be readily obtained. The detection sensitivity is increased by about ten times and mass resolution is increased to about 200 fwhm. However, as compared with static MALDI-TOF experiments, this resolution is still rather low.

Another interface has been developed by Zhan et al. [95] which should allow the direct coupling of capillary LC with MALDI-TOF. The interface employs continuous analyte/matrix co-crystallization onto a porous frit installed at a capillary end which is used as the target for MALDI. After separation, the analyte effluent is premixed with the MALDI matrix solution and introduced into the interface. The analyte/matrix mixture is co-crystallized onto the frit surface in the vacuum environment of the mass spectrometer. Continuous matrix/analyte crystallization and interface regeneration is accomplished by a combination of solvent flushing and laser ablation. Several applications, including small peptides and low molar mass polyethylene glycol, have been given.

References
1. BALOGH MP (1997) LC-GC Int 10:728
2. SMITS R (1995) LC-GC Int 8:92
3. NIESSEN WMA, TINKE AP (1995) J Chromatogr A703:37
4. DOLE M (1968) J Chem Phys 49:2240
5. HORNING EC (1974) J Chromat Sci 12:725
6. VESTAL ML, FERGUSON GJ (1985) Anal Chem 57:2373
7. WILLOUGHBY RC, BROWNER RF (1984) Anal Chem 56:2626
8. YERGEY AL, EDMONDS CG, LEWIS IAS, VESTAL ML (1990) Liquid chromatography/mass spectrometry - techniques and applications. Plenum Press, New York
9. ARPINO PJ (1990) Mass Spectrom Rev 9:631
10. ARPINO PJ (1992) Mass Spectrom Rev 11:3
11. BEHYMER TD, BELLAR TA, BUDDE WL (1990) Anal Chem 62:1686
12. VOYKSNER RD, SMITH CS, KNOX PC (1990) Biomed Environ Mass Spectrom 19:523
13. SCIEX PE (1989) The API Book, Thornhill, Ontario
14. WHITEHOUSE CM, DREYER RN, YAMASHITA M, FENN JB (1985) Anal Chem 57:675
15. FENN JB, MANN M, MENG CK, WONG SF, WHITEHOUSE CM (1990) Mass Spectrom Rev 9:37
16. BRUINS AP, COVEY TR, HENION JD (1987) Anal Chem 59:2642
17. VOUROS P, WRONKA JW (1991) In: Barth HG, Mays JW (eds) Modern methods in polymer characterization. Wiley, New York, , Chap. 12
18. VESTAL ML (1984) Science 226:595
19. VARGO JD, OLSON KL (1985) Anal Chem 57:672
20. VESTAL ML (1983) Mass Spectrom Rev 2:447
21. VARGO JD, OLSON KL (1986) J Chromatogr 353:215
22. PROKAI L, SIMONSICK WJ (1995) SEC with electrospray mass spectrometric detection. In: Provder T, Barth HG, Urban MW (eds) Chromatographic characterization of polymers. Hyphenated and multidimensional techniques. Adv Chem Ser 247, American Chemical Society, Washington, DC, Chapter 4
23. PROKAI L, SIMONSICK WJ (1993) Rapid Commun Mass Spectrom 7:853
24. SIMONSICK WJ (1993) Polym Mater Sci Eng 69:412
25. SIMONSICK WJ, ROSS CW (1996) Polym Prepr 37:286
26. PROKAI L, MYUNG SW, SIMONSICK WJ (1996) Polym Prepr 37:288

27. KEMP TJ, BARTON Z, MAHON A (1996) Polym Prepr 37:305
28. PASCH H, RODE K (1995) J Chromatogr A699:21
29. DANIS PO, SAUCY DA, HUBY FJ (1996) Polym Prepr 37:311
30. DANIS PO, HUBY FJ (1995) J Am Soc Mass Spectrom 6:1112
31. MONTAUDO G, GAROZZO D, MONTAUDO MS, PUGLISI C, SAMPERI F (1995) Macromolecules 28:7983
32. MONTAUDO MS, PUGLISI C, SAMPERI F, MONTAUDO G (1998) Macromolecules 31:3839
33. PASCH H, TRATHNIGG B (1998) HPLC of polymers. Springer, Berlin Heidelberg New York
34. PASCH H, BRINKMANN C, MUCH H, JUST U (1992) J Chromatogr 623:315
35. GORSHKOV AV, MUCH H, BECKER H, PASCH H, EVREINOV VV, ENTELIS SG (1990) J Chromatogr 523:91
36. GORSHKOV AV, VERENICH SS, EVREINOV VV, ENTELIS SG (1988) Chromatographia 26:338
37. ADRIAN J, BRAUN D, RODE K, PASCH H (1999) Angew Makromol Chem 267:73
38. LEE H, LEE W, CHANG T, CHOI S, LEE D, JI H, NONIDEZ WK, MAYS JW (1999) Macromolecules 32:4143
39. PASCH H, BRINKMANN C, GALLOT Y (1993) Polymer 34:4099
40. PASCH H, AUGENSTEIN M (1993) Makromol Chem 194:2533
41. BRAUN D, ESSER E, PASCH H (1998) Int J Polym Char 4:501
42. PASCH H (1997) Adv Polym Sci 128:1
43. KRÜGER RP, MUCH H, SCHULZ G (1996) GIT Fachz Lab 4:398
44. JUST U, KRÜGER RP (1996) In: Auner N, Weis J (eds) Organosilicon chemistry II, VCH Weinheim
45. MONTAUDO G (1995) Rapid Commun Mass Spectrom 9:1158
46. MONTAUDO G (1998) Proc 11th Int Symp Polym Anal Char, Santa Margerita Ligure, Italy
47. KRÜGER R-P, MUCH H, SCHULZ G (1996) Int J Polym Anal Charact 2:221
48. KOTTNER N, BUBLITZ R, KLEMM E (1996) Macromol Chem Phys 197:2665
49. RÄDER H-J, SPICKERMANN J, KREYENSCHMIDT M, MÜLLEN K (1996) Macromol Chem Phys 197:3285
50. SPICKERMANN J, RÄDER H-J, MÜLLEN K, MÜLLER B (1996) Macromol Rapid Commun 17:885
51. MONTAUDO G (1996) Polym Prepr ACS Div Polym Chem 37:290
52. MONTAUDO G (1996) TRIP 4:81
53. BRAUN D, HENZE I, PASCH H (1997) Macromol Chem Phys 198:3365
54. PASCH H, DEFFIEUX A, GHAHARY R, SCHAPACHER M, RIQUE-LURBET L (1997) Macromolecules 30:98

55. Wachsen O, Reichert KH, Krüger RP, Much H, Schulz G (1997) Polym Degrad Stab 55:225
56. Lo T-Y, Huang SK (1998) J Appl Polym Sci 68:1621
57. Hunt SM, Derrick PJ, Sheil MM (1998) Eur Mass Spectrom 4:475
58. Montaudo MS (1999) Macromol Symp 141:95
59. Ihara E, Tanabe M, Nakayama Y, Nakamura A, Yasuda H (1999) Macromol Chem Phys 200:758
60. Montag P (1999) CLB Chemie in Labor und Biotechnik 50:253
61. Nielen MWF, Malucha S (1997) Rapid Commun Mass Spectrom 11:1194
62. Montaudo MS, Puglisi C, Samperi F, Montaudo G (1998) Rapid Commun Mass Spectrom 12:519
63. Gagel JJ, Biemann K (1986) Anal Chem 58:2184
64. Gagel JJ, Biemann K (1987) Anal Chem 59:1266
65. Wheeler LM, Willis JN (1993) Appl Spectrosc 47:1128
66. Willis JN, Dwyer JL, Wheeler LM (1993) Polym Mat Sci 69:120
67. Pasch H, Esser E, Montag P (1996) GIT Fachz Lab Chromatogr 16:68
68. Willis JN, Wheeler L (1995) Use of a GPC-FTIR interface for polymer analysis. In: Provder T, Barth HG, Urban MW (eds) Chromatographic characterization of polymers. Hyphenated and multidimensional techniques. Adv Chem Ser 247, American Chemical Society, Washington, DC, Chapter 19
69. Chervet J-P, van den Broek R (1997) GIT Lab J Int Ed 1:92
70. Product documentation Probot, bai GmbH, Lautertal, Germany
71. Liedke S, Kremser L, Allmaier G, Rizzi A, Roitinger A (2001) GIT BIOforum 24:152
72. Esser E, Keil C, Braun D, Montag P, Pasch H (2000) Polymer 41:4039
73. Falkenhagen J, Krüger R-P, Schulz G, Gloede J (2001) GIT Fachz Lab 45:380
74. Böhmer V (1995) Angew Chem 107:785
75. Gutsche CD (1998) Calixarenes revisited. In: Stoddart JF (ed) Monographs in supramolecular chemistry. Royal Society of Chemistry, Cambridge
76. Nielen MWF (1998) Anal Chem 70:1563
77. Nielen MWF, Buijtenhuijs FA (2001) LC-GC Europe 14:82
78. Nielen MWF, Malucha S (1997) Rapid Commun Mass Spectrom 11:1194
79. Keil C, Esser E, Pasch H (2001) Macromol Mat Eng 286:161
80. Li L, Wang APL, Coulson LD (1993) Anal Chem 65:493
81. Caprioli RM (1990) Anal Chem 62:477A

82. Murray KK, Russell DH (1993) Anal Chem 65:2534
83. Murray KK, Lewis TM, Beeson MD, Russell DH (1994) Anal Chem 66:1601
84. Murray KK, Russell DH (1994) J Am Soc Mass Spectrom 5:1
85. Fei X, Wei G, Murray KK (1996) Anal Chem 68:1143
86. Fei X, Murray KK (1996) Anal Chem 68:3555
87. Nagra DS, Li L (1995) J Chromatogr A711:235
88. Coulson LD, Nagra DS, Guo X, Whittal RM, Li L (1994) Appl Spectrosc 48:1125
89. Whittal RM, Russon LM, Li L (1998) J Chromatogr A794:367
90. Joos PE (1995) LC-GC Int 8:92
91. Montaudo MS, Puglisi C, Samperi F, Montaudo G (1998) Rapid Commun Mass Spectrom 12:519
92. Marziarz EP, Liu XM, Quinn ET, Lai YC, Ammon DM, Grobe GL (2002) J Am Soc Mass Spectrom 13:170
93. Kassis CE, DeSimone JM, Linton RW, Remsen EE, Lange GW, Friedman RM (1997) Rapid Commun Mass Spectrom 11:1134
94. Dwyer J, Botten D (1997) Internat Lab 27:13A
95. Zhan Q, Gusev A, Hercules DM (1999) Rapid Commun Mass Spectrom 13:2278
96. Liu J, Loewe RS, McCullough RD (1999) Macromolecules 32:5777
97. Montaudo MS, Montaudo G (1999) Macromolecules 32:7015
98. Montaudo MS (2001) Macromolecules 34:2792
99. Mengerink Y, Peters R, deKoster CG, van der Wal S, Claessens HA, Cramers CA (2001) J Chromatogr A914:131
100. Nielen MWF (1999) Mass Spectrom Rev 18:309
101. Gusev AI (2000) Fresenius J Anal Chem 366:691

8 Recent Developments and Outlook

This chapter describes some recent developments and chances for the future of MALDI MS. As already pointed out, the actual main driving force for technical advancements is the area of proteomics [1]. The requirements of high-throughput screening systems lead to considerable progress in automation, sample preparation, and interpretation. Without doubt the past decade has also seen a vivid development of MALDI applications to synthetic polymers but also limitations especially concerning polymers with broad molar mass distributions. Nevertheless nearly all items of MALDI MS are still under development.

First let us consider the field of matrices, sample preparation, and ionization procedure. The search for new matrices, as a key parameter of successful MALDI measurements, is continuing. With 4-hydroxybenzylidene malonitrile (HBM) a matrix has been discovered which produces less fragments and adducts in the low molar mass region, whereas 2-(2E)-3-(4-tert-butylphenyl)-2-methyl-2-enylidene]malonitrile (DCTB) allows the recording of positive and negative ion spectra at lower threshold laser fluences, thus giving less unwanted fragments [2]. A broad range of analytes (proteins, cyclodextrins, PPGs) is reported to give good positive and negative ion spectra in 5-ethyl-2-mercaptothiazole [3].

Also new aspects concerning solvent-free sample preparation as a more universal method are reported. The recording of spectra of PS, PMMA, and polyetherimide in a mass range between 2 and 100 kDa is described by Räder et al. [4]. Spectra of otherwise insoluble polycyclic aromatic hydrocarbons (PAHs) have been obtained by mixing with the matrix 7,7,8,8-tetra-cyanochinodimethane without any solubilization procedures [5]. These mixing procedures should be of advantage in general in cases of hardly dissolvable samples like industrial pigments and broaden the capabilities of MALDI MS. It has been shown that there is fundamentally no difference in the quality of the obtained spectra for the solvent-based and the solvent-free preparation procedures. It was, therefore, concluded that the mechanism of desorption and ionization remains unchanged.

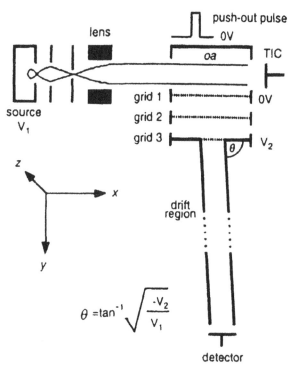

Fig. 8.1. Scheme of the oa-TOF analyzer, TIC = total ion current detector. (Reprinted from [6] with permission of Wiley, UK)

Considerable progress is reported for the analyzer design. Here first of all the so-called orthogonal acceleration (oa)-TOF analyzer has to be mentioned. In this arrangement [6], ion source, TOF analyzer, and detector describe an angle slightly larger than 90°. A low energy collimated ion beam fills an acceleration region. One or more electrodes in the accelerator are pulsed on to create an extraction field orthogonal to the continuous ion beam. The ions deflected by the push-out pulse are then accelerated into the TOF analyzer. The timing is arranged in such a way that ions for the next cycle of extraction are accumulating in the accelerator region while ions of the current cycle are drifting to the detector.

This method of ion gating is very efficient. It allows the use of continuous ion sources, i.e., the combination of TOF-analyzers with ESI sources which in turn can be coupled to other continuous separation techniques like liquid chromatography. Further on, the collimation of the ion beam produces ions with a minimal velocity spread in the direction of the TOF analyzer resulting in an improvement in mass resolution. A considerable advantage arises by the fact that with oa-TOF analyzers MS/MS experiments can be readily realized (the PSD technique described above causes

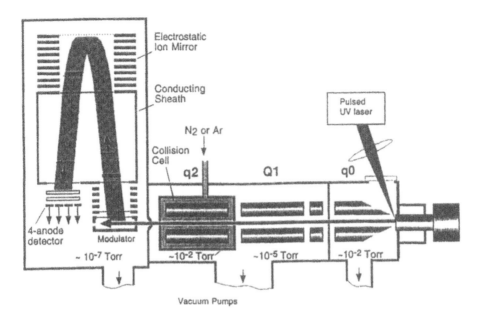

Fig. 8.2. Schematic diagram of the MALDI-Qq-TOF mass spectrometer. The resulting pulsed ion beam is converted into a quasi-continuous beam by collisonal cooling in q0. In q2 collision induced dissociation (CID) is achieved. (Reprinted from [9] with permission of Wiley, UK)

a loss in sensitivity and resolution). Particularly successful are hybrid quadrupole Q-TOF mass spectrometers [7,8]. Here the parent ion is selected in a quadrupole mass filter Q1, is fragmented in a collision cell within a RF-only ion guide (see Sect. 2.2.1). The daughter ions then enter a reflecting TOF analyzer, where all ions are detected in parallel. This combination offers high sensitivity, high mass resolution and accuracy and fast daughter ion analysis, facts which make it more convenient than for example triple quadrupole instruments.

The goal to achieve MS and MS/MS experiments with high accuracy from the same sample on the same device has been achieved recently with a MALDI-Q-TOF instrument [10]. The coupling to the MALDI source is accomplished by an RF quadrupole working at a pressure above a few mTorr. Hereby collisional cooling of the ions is achieved and the pulsed MALDI beam is transformed into a quasi-continuous beam [11], i.e., the desorption process is almost decoupled from the analysis. Also the application in tandem mass spectrometers making use of collision induced dissociation is reported [12]. Here an individual oligomeric ion is selected from a MALDI generated ion pulse by means of a magnetic sector instrument and directed through a collision region. An example for the analysis of PS is shown in Fig. 8.3.

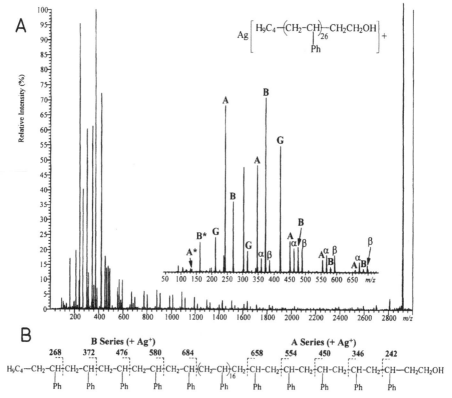

Fig. 8.3. **A** MALDI-CID spectrum of the parent ion 2914.6 amu of the 26-mer of polystyrene given in the inset (M + Ag]$^+$), structure of the parent ion and enlarged section of the spectrum. **B** Proposed fragmentation pathway of the parent ion showing how the A and B series may be used to infer the respective endgroup. Reprinted from [12] with permission of Elsevier, The Netherlands)

The two main series of the parent ion at 2914.6 amu are explained in the illustration, allowing the determination of the different endgroups. Most of the fragments retain the cation (Ag$^+$ in this case). The other series (G, α, β) showing up in the inset of Fig. 8.3 can be explained as the result of rearrangement reactions. The upper limit for the MALDI-CID experiment at present is around 6 kDa.

Further attempts to get beyond PSD are reported. An interesting approach to achieve MS^n-capability in MALDI MS is reported by Tanaka et al. [13]. A quadrupole ion trap (QIT) is coupled to a TOF analyzer. Claimed advantages of this set-up are high mass accuracy, a wide mass range (up to 50 kDa), high parent ion resolution, high sensitivity due to optimized ion optics, and high trapping efficiency. Mass resolution and accuracy are independent of the ionization conditions and maintained at any MS^n-level.

Other approaches try to reduce the time necessary to record PSD spectra (the stepwise change in the reflectron voltage). One is to use a timed ion gate and a quadratic field reflectron (Z^2 technology) allowing a single-run recording of a complete PSD spectrum [14]. A complete spectrum with low sample consumption can be recorded in 1–2 min compared to 20–30 min in a conventional MALDI-TOF instrument. In a second approach [15] the parent ions and fragments are "lifted" to a higher potential. This method also allows the recording of PSD spectra in a single run without sacrificing the performance. This highly efficient procedure corresponds to a tandem mass spectrometer combining two TOF analyzers (MALDI-TOF/TOF).

The use of MALDI sources in FT-CIR instruments promises a drastic improvement in mass resolution: 65,000 for insulin and 85,000 for the 146-mer of PEG have been reported 16].

The (on-line) coupling of MALDI MS to other (pre)separation techniques still is an active field. Combination with SEC has been described in Chap. 7 and in [17]. An aerosol MALDI spectrometer allowing a direct liquid introduction is described in [18]. An atmospheric pressure MALDI/ion trap mass spectrometer is described in [19]. This configuration enables the application-specific selection of the appropriate ionization source, for example electrospray/atmospheric pressure chemical ionization or MALDI. The detection limit in the MALDI mode for peptide analytes is between 10 and 50 fmol and MS/MS experiments could be performed.

Of course, with the development of new synthetic approaches the range of applications of MALDI MS still broadens. Recent examples include the investigation of low bandgap conjugated polymers for solar cells [20], of conjugated polyrotaxanes for insulated molecular wires [21], of hyperbranched polyglyceroles [22], of nanocomposites by surface-initiated living cationic polymerization of oxazolines [23], of the biodegradation of intermediates of non-ionic surfactants [24], or applications to amphiphilic fullerenes [2].

From this arbitrary collection of newer developments one must conclude that MALDI MS of synthetic polymers still is an evolving technique and in the future will experience further progress.

References

1. WILKINS MR, WILLIAMS KL, APPEL RD, HOCHSTRASSER DF (Eds.) (1997) Proteome research: new frontiers in functional genomics. Springer, Berlin Heidelberg New York
2. BROWN T, CLIPSTON NC, SIMJEE N, LUFTMANN H, HUNGERBÜHLER H, DREWELLO T (2001) Int J Mass Spectrom 210/211:249

3. Rajin NP, Mirza SP, Vairamani M, Ramulu AR, Pardhasaradhi M (2001) Rapid Commun Mass Spectrom 15:1884
4. Trimpin S, Rouhanipour A, Az R, Räder HJ, Müllen K (2001) Rapid Commun Mass Spectrom 15:1364
5. Przybilla L, Brand JD, Yoshimura K, Räder HJ, Müllen K (2000) Anal Chem 72:4597
6. Guilhaus M (1995) J Mass Spectrom 30:1519
7. Morris HR, Paxton T, Dell A, Langhorne J, Berg M, Bordoli R, Hoyes J, Bateman RM (1996) Rapid Commun Mass Spectrom 10:889
8. Shevchenko A, Chernushevich IV, Ens W, Standing KG, Thomson B, Wilm M, Mann M (1997) Rapid Commun Mass Spectrom 11:1015
9. Loboda AV, Krutchinsky AN, Bromirski M, Ens W, Standing KG (2000) Rapid Commun Mass Spectrom 14:1047
10. Shevchenko A, Loboda A, Ens W, Standing KG (2000) Anal Chem 72:2132
11. Krutchinsky AN, Loboda AV, Spicer VL, Dworschak R, Ens W, Standing KG (1998) Rapid Commun Mass Spectrom 12:50
12. Jackson AT, Bunn A, Hutchings LR, Kiff FT, Richards RW, Williams J, Green MR, Bateman RH (2000) Polymer 41:7437
13. Tanaka K, Kawatoh E, Ding L, Smith AJ, Kumashiro S (1999) Proc. 47th ASMS Conf Mass Spectrom Allied Topics, TP 086
14. Keough T (1999) Proc Nat Acad Sci 96:7131
15. Franzen J, Frey R, Holle A, Krauter O (2001) Int J Mass Spectrom 206:275
16. Easterling ML, Pitsenberger CC, Kulkarni SS, Taylor PK, Amster IJ (1996) Int J Mass Spectrom Ion Processes 157/158:97
17. Hanton SD, Liu XM (2000) Anal Chem 72:4554
18. Fei X, Wei G, Murray KM (1996) Anal Chem 68:1143
19. Laiko VV, Moyer SC, Cotter RJ (2000) Anal Chem 72:5239
20. Dhanabalan A, VanDuren JKJ, VanHul PA, van Dongen JLJ, Janssen RAJ (2001) Adv Funct Mater 11:255
21. Taylor PN, O'Connell MJ, McNeill LA, Hall M, Aplin RT, Anderson (2000) Angew Chem 39:3460
22. Kautz H, Sunder A, Frey H (2001) Macromol Symp 163:67
23. Jordan R, West N, Ulman A, Chou YM, Nuyken O (2001) Macromolecules 34:1606
24. Sato H, Shibata A, Wang Y, Yoshida H, Tamura H (2001) Polym Degr Stab 74:69

Subject Index

A
acceleration voltage 34
aerosol MALDI-TOF 280
aerosol method 279
air-spray deposition 66
amphiphilic fullerenes 293
analytical ultracentrifugation 9
atmospheric pressure chemical ionisation 209
atmospheric pressure ionization 210
atom transfer radical polymerization 257

B
bisphenol-A polycarbonate 50, 89
block copolymers 246, 257
block copolymers of ethylene oxide and propylene oxide 230
block copolymers of L-lactide and ethylene oxide 246
branching 244

C
calibration 238
calibration lines 220
calixarenes 262
caprolactam-pyrrolidone copolymer 100
cationization 109
cationizing agent 71
–, acid/base concept 73
–, endgroups 73
–, quantitative analysis 73
chemical composition 135
chemical heterogeneity 3, 11, 68
–, particle size distribution 11
chemical structure 135
cis-1,4-polyisoprene 115
collision induced dissociation 42
complex polymers 135
condensation polymers 191
conformation 12

continuous-flow (cf) MALDI-TOF 279
copolymer 269
copolymer of vinyl acetate and vinyl pyrrolidone 215
copolymers 99, 169, 176, 203
–, average sequence lengths 169
–, isotope separation 180
–, random copolymer 169
copolymers of ethylene oxide and propylene oxide 176
coupling of liquid chromatography and MALDI-TOF MS 209, 252
coupling of MS techniques 42

D
Dalton 19
data processing 40
–, Ion counting 41
degree of polymerization 107
delayed ion extraction 75, 135
detectors 38
–, Daly detector 39
–, electron multiplier 39
–, Faraday cup 39
–, focal plane detector 39
–, multichannel plates 39
–, scintillator 39
determination of endgroups 136
determination of molar masses 50
diblock copolymers of α-methylstyrene and 4-vinylpyridine 170
direct coupling of liquid chromatography and MALDI-TOF MS 279
dried droplet method 65

E
electron impact ion (EI) sources 22
–, electron abstraction 22
electrospray 210
electrospray deposition 66

electrospray ionization (ESI)   23, 209
–, mass spectrometry   50
emulsifiers   96
endgroup analysis   168
endgroup functionality   191
endgroups   102, 228
–, fragment ion   102, 103
–, ion   102
–, mother ion   103
–, parent   102
energy focusing   36
energy pooling   60
epoxy resins   235
–, functionality type distribution   236
–, molar mass distribution   236
ESI-MS   212
excited-state proton transfer   60

F
fast atom bombardment   25
fatty alcohol ethoxylates   224
–, calibration curves   229
field desorption   25
field desorption mass spectrometry   44
field ionization   25
fragmentation   21, 42
functional heterogeneity   161, 213
functionality   135, 155, 238
functionality type distribution   3, 136, 269
–, number-average functionality   137
–, weight-average functionality   137

G
gas phase cationization   62
gas phase ionization   22
gas phase proton transfer   61
–, proton affinities   61
GC-MS   209
gel permeation chromatography   107
glycerol-started PPO   270
group transfer   257
group-transfer polymerization   154

H
heterogeneity   246
homopolymers   89
HPLC-MS coupling   209
hydrocarbon polymers   115
hyperbranched polyglyceroles   293
hyphenation   44
hyphenation of MS   42

I
identification of polymers   85
–, repeat unit   85
interfaces for coupling LC and MALDI-TOF MS   252
interpretation of spectra   67
–, mass calibration   67
ion gating   290
ion sources   21
ion velocity   34
ionization   58
–, initial velocity spread   59
–, primary ion formation   59
–, time-delayed extraction   59
ionization from the condensed phase   24
ionization mechanism   60
ionization potentials   60

L
laser desorption   28, 45
–, resonance enhanced multiphoton ionization   28
laser desorption/ionization   57
laser power   73, 79, 128
–, desorption/ionization   79
–, threshold   79
–, threshold value   73
LC transform interface   252
–, matrix-precoated foils   254
–, SEC separation   254
LC-CC   240, 246
–, HPLC fractionation   247
LC-CC-MALDI-TOF   262, 269
LC-MS interfaces   210, 211
light scattering   7
–, Brillouin scattering   7
–, Doppler effect   7
–, dynamic light scattering   9
–, hydrodynamic radius   9
–, radius of gyration   8
–, Rayleigh scattering   7
–, scattering vector   7
–, small angle light scattering ($\theta = 6°$)   8
–, static light scattering   7
–, wide angle light scattering   8
–, Zimm diagram   8
liquid adsorption chromatography   15
–, calibration curve   16
liquid chromatography   15
liquid chromatography at the critical point of adsorption   224, 230

Subject Index

M

μSEC-MALDI-TOF 266
macromonomers 136
MALDI MS 57
MALDI process 57, 58
MALDI-Q-TOF 291
MALDI-TOF interface 256
-, measurement of LC fractions 256
MALDI-TOF mass spectrometry 57, 108, 135
-, endgroups 108
-, repeat units 108
MALDI-TOF MS 250
-, off-line detector 250
Mark-Houwink coefficients 12
mass 75
mass analyzers 19, 29
-, collision induced dissociation 32
-, fourier transform analyzer 32
-, ion trap 32
-, quadrupole analyzer 29
-, stability diagram 31
-, time-of-flight analyzer 34
mass resolution 19, 35
mass spectrometry 19, 57
-, atomic mass 19
-, atomic mass units 19
matrix 57, 60, 62–64, 70, 109, 110, 289
-, cationizing agents 63
matrix clusters 59
-, multiphoton ionization 59
metastable ions 42
methods for molar mass determination 6
microcolumn LC-MALDI system 282
microcolumn liquid chromatography 255
microfractionation 255
modified silicone copolymers 180
molar mass 2, 107, 115, 118, 122, 135, 238
molar mass determination 107, 132
molar mass distribution 2, 3, 27, 78, 107, 112, 219, 221
molecular architectures 3
molecular heterogeneity 2, 4, 85, 135, 191
-, number-average molar mass 4
-, weight average molar mass 4
-, z-average of molar mass 5
most probable peak value 110
MS/MS experiments 290
MS/MS instrumentation 43

N

nanocomposites 293
narrow molar mass distribution 112
number-average molar mass 107

O

off-line LC separations 212
oligo(caprolactone) 213, 214
oligomer 112
on-line coupling 279
orthogonal acceleration (oa)-TOF analyzer 290

P

particle beam ionization 209
PEG 80, 121
perfluorinated polyether 48, 50
phenol-formaldehyde resins 191
phenolic novolacs 89, 192
phenolic resins 191
phenolic resol 195
phenol-urea-formaldehyde oligomers 197
PMMA 74, 75, 77, 81, 121, 131, 154
-, GTP polymerized 154
poly(α-methylstyrne)-b-poly(4-vinylpyridine) diblock copolymers 170
poly-ε-caprolactone 162
-, isotopic distribution 166
poly(dimethylsiloxane) 51, 52, 97, 180
poly(ethylene oxide)-b-poly(L-lactide)s 247
poly(ethylene oxides) 222
poly(ethyleneoxide)-b-poly(p-phenylene ethynylene) 103
poly(methylphenyl silane) 150
poly(n-butyl methacrylate-$block$-methyl methacrylate) 257
poly(oxymethylene) 95
poly(tetramethylene ethylene glycol) 110, 111
polyamide-6 82, 122, 138
polyamides 138
polybutadiene 115, 120
polycarbonate 122, 266
polydispersity 5, 78, 107, 108, 113, 118, 121 pp, 126, 219
polyester copolymers 219
polyester diols 93
polyesters 122
polyethylene glycol 80, 113, 123, 279
-, desorption/ionization 124
-, laser powers 124

polyethylene terephthalate 89
polyglycerolester 97
polyisoprene 120
polymethyl methacrylate 113, 126, 130, 215
polyolefins 25
polypropylene oxides 269
polyrotaxanes 293
polysiloxane 97
polystyrene 58, 76, 113, 115pp, 119, 121, 126, 130, 186, 292
polystyrene sulfonic acids 186
polysulfones 90
polyurethanes 49, 198, 230
-, hard segments 200
-, selective degradation 200
polyvinyl acetate 215
post source decay 102
PROBOT interface 255
propylene oxide 269
proton transfer 60
-, secondary ionization 61
PS 68, 76, 78, 130
-, endgroups 69
PSD 292
pyrolysis MS 46

Q
quadrupole ion trap 292
-, MALDI-TOF/TOF 293
-, Z^2 technology 293
quadrupole Q-TOF mass spectrometers 291

R
radius of gyration 17
reflectron 36, 37, 75
resolution 20, 75
-, endgroup 75

S
sample preparation 62, 65, 67, 109
-, cationization agent 62
-, ionization 62

SEC 113, 160, 161, 217, 219, 238
-, cationizing agents 113
SEC-MALDI-TOF 269
secondary ion mass spectrometry 25
-, collision cascade 26
secondary ion mass spectrometry 48
sensitivity 19, 20
signal-to-noise ratio 20
size exclusion chromatography 13, 107, 214
-, calibration curve 14
-, calibration of 214
-, distribution coefficient 13
-, Gibbs free energy 14
-, hydrodynamic volume 13
-, rentention volume 13
soft ionization 58
solvent-free sample preparation 289
spin-coating 66
spray techniques 22
static SIMS mode 27
sulfonation of polystyrene 186

T
tandem mass spectrometry 43
telechelics 136
thermospray ionization 209
time-lag focusing 37, 75
TOF analyzer 73
triazine-based polyamines 143
triblock copolymers 230

U
ultracentrifuge 9
-, density gradient run 10
-, sedimentation equilibrium run 10
-, sedimentation velocity run 10

V
viscosity 12
viscometry 12

W
weight-average molar mass 107

Printing (Computer to Plate): Saladruck Berlin
Binding: Stürtz AG, Würzburg

Lightning Source UK Ltd.
Milton Keynes UK
UKOW06n1636040915

257941UK00009BE/60/P